Eva Baake

Studie zur Ursache und Therapie der equinen rezidivierenden Uveitis

Cuvillier Verlag Göttingen
Internationaler wissenschaftlicher Fachverlag

Bibliografische Information der Deutschen Nationalbibliothek

Die Deutsche Nationalbibliothek verzeichnet diese Publikation in der Deutschen Nationalbibliografie; detaillierte bibliografische Daten sind im Internet über http://dnb.d-nb.de abrufbar.

1. Aufl. - Göttingen: Cuvillier, 2017

Zugl.: Hannover (TiHo), Univ., Diss., 2017

Nonnenstieg 8, 37075 Göttingen

Telefon: 0551-54724-0

Telefax: 0551-54724-21

www.cuvillier.de

1. Auflage, 2017

Gedruckt auf umweltfreundlichem, säurefreiem Papier aus nachhaltiger Forstwirtschaft.

ISBN 978-3-7369-9651-9

eISBN 978-3-7369-8651-0

Tierärztliche Hochschule Hannover

Studie zur Ursache und Therapie der equinen rezidivierenden Uveitis

INAUGURAL-DISSERTATION

zur Erlangung des Grades einer Doktorin der Veterinärmedizin

- Doctor medicinae veterinariae -

(Dr. med. vet.)

vorgelegt von

Eva Ingrid Agnes Baake, geb. Nordemann

Thuine

Hannover 2017

Wissenschaftliche Betreuung: Prof. Dr. Bernhard Ohnesorge

Klinik für Pferde

Tierärztliche Hochschule Hannover

1. Gutachter: Prof. Dr. B. Ohnesorge, Klinik für Pferde TiHo Hannover

2. Gutachter: Prof. Dr. P. Valentin-Weigand, Institut für Mikrobiologie TiHo Hannover

Tag der mündlichen Prüfung: 25.10.2017

Für Konstantin

Teile der Dissertation wurden unter folgendem Titel veröffentlicht:

Detection of intraocular leptospiral DNA, antibodies and Leptospira spp. in horses with equine recurrent uveitis in different laboratories

Intraokularer Leptospirennachweis mittels PCR, MAT, ELISA und Kultur von Pferden mit equine rezidivierender Uveitis (ERU) in verschiedenen Laboren

E. Baake, M. von Borstel, K. Rohn, B. Ohnesorge

Pferdeheilkunde, 32, 4 (2016), S. 346-356

URL: http://www.hippiatrika.com/download.htm?id=20160407

DOI: 10.21836/PEM20160407

Ergebnisse der Dissertation wurden in Form eines Posters auf folgender Fachtagung präsentiert:

DVG-Vet-Congress 2016 in Berlin

Detection of intraocular leptospiral DNA, antibodies and Leptospira spp. in horses with equine recurrent uveitis in different laboratories

E. Baake, M. von Borstel, K. Rohn, B. Ohnesorge

Inhalt

Abkürzungsverzeichnis

ERU	Equine Rezidivierende Uveitis
MAT	Mikroagglutinationstest
PCR	Polymerasekettenreaktion
ELISA	Enzyme-linked Immunosorbent-Assay
L1	Labor 1
L2	Labor 2
L3	Labor 3
REC	recurrence of uveitis
NOIS	non-specific ocular inflammatory symptoms
FOC	free of complaints
IMP	improvement

1 Einleitung

Die equine rezdivierende Uveitis (ERU) ist eine Augenerkrankung entzündlichen Charakters, die durch ihren progredienten und wiederkehrenden Verlauf signifikante Auswirkungen auf die betroffenen Pferde sowie ihre Besitzer hat (Gerding & Gilger, 2016). Im Wesentlichen werden drei verschiedene Formen der ERU anhand der betroffenen Augenabschnitte unterschieden. Erstens die schmerzhafte „anteriore Uveitis/Iritis“, zweitens die wenig bis nicht schmerzhafte „intermediäre Uveitis/Zyklitis“, welche auch „schleichende Uveitis“ genannt wird und drittens die „posteriore Uveitis/Choroiditis“. Diese drei Formen können sowohl isoliert, als auch gemeinsam auftreten und werden dann auch als „Panuveitis“ bezeichnet (Spiess, 2010; Tömördy et al, 2010; Wollanke & Gerhards, 2009).

Es hat sich gezeigt, dass einige Formen der ERU häufiger bei bestimmten Pferderassen diagnostiziert werden. So wird die „intermediäre Uveitis/Zyklitis“ häufig bei Appaloosas und Kaltblutpferden gesehen (Gilger, 2010). Die „anteriore Uveitis/Iritis“ tritt häufig zusammen mit der „posterioren Uveitis/Choroiditis“ auf und wird bisweilen als „Klassische Uveitis“ bezeichnet (Gilger, 2010).

Immer wieder wird die ERU mit einer intraokularen Leptospireninfektion in Verbindung gebracht (Ahmed et al, 2006; Alexander, 1990; Brandes et al, 2007; Brem et al, 1998; 1999; Faber et al, 2000; Frellstedt, 2009; Gerding & Gilger, 2016; Halliwell, 1985; Hartskeerl et al, 2004; Heusser, 1948; Polle et al, 2014; Rimpau, 1947; Schwink et al, 1989; Sillerud et al, 1987; Von Borstel et al, 2010; Wiehen, 2012; Wollanke et al, 1998; Wollanke et al, 2004; Wollanke et al, 2000; Wollanke et al, 2001). Die intraokulare Leptospireninfektion kann bestätigt werden durch eine erfolgreiche Anzucht der Bakterien, durch einen Antigennachweis, oder durch einen

Antikörpernachweis aus Kammerwasser bzw. Glaskörpermaterial von an ERU erkrankten Pferden. Leptospirales Antigen kann mit Hilfe der Polymerasekettenreaktion (PCR), und leptospirale Antikörper können mit Hilfe des Mikroagglutinationstestes (MAT) nachgewiesen werden. Die positiven Nachweisraten der Antikörper mittels MAT schwanken in der Literatur zwischen 43 und 94 % (Brandes et al, 2007; Brem et al, 1999; Gerding & Gilger, 2016; Gilger et al, 2008; Polle et al, 2014; Von Borstel et al, 2010; Wollanke et al, 1998; Wollanke et al, 2001). Die PCR kann bei 40 bis 100 % der betroffenen Pferde leptospirales Antigen aus Kammerwasser- beziehungsweise aus Glaskörpermaterial nachweisen (Brandes et al, 2007; Faber et al, 2000; Polle et al, 2014; Von Borstel et al, 2010), und lebende Leptospiren können laut Literatur in 16 bis 53 % der Proben erfolgreich angezüchtet werden (Brandes et al, 2007; Brem et al, 1999; Dorrego-Keiter E, 2016; Faber et al, 2000; Hartskeerl et al, 2004; Wollanke et al, 2004; Wollanke et al, 2001). Da in einem nicht zu vernachlässigendem Anteil der Fälle keine Hinweise auf eine intraokulare Leptrospireninfektion vorliegen, ist davon auszugehen, dass es grundsätzlich eine leptospiren-positive und eine lepstospiren-negative Form der ERU gibt (Gerding & Gilger, 2016).

Der Austausch von Glaskörpermaterial im Rahmen einer Vitrektomie der an ERU erkrankten Pferdeaugen wird nun seit mehr als zwei Jahrzehnten in Europa durchgeführt. Es zeigte sich in den ersten Studien, dass die Vitrektomie durch den Austausch von entzündlich verändertem, trüben Glaskörpermaterial zu einer Aufklarung des Glaskörperraumes führt und der progrediente Verlauf der Erkrankung auf diese Weise positiv beeinflusst werden kann (Werry & Gerhards, 1992; Werry, 1991). Weitere Studien aus den darauffolgenden Jahren zeigten, dass zwischen 73,6 und 100 % der vitrektomierten Pferde keine rezidivierenden Uveitis-Schübe auf dem operierten Auge zeigten (Frühauf, 1998; Tömördy et al, 2010; Von Borstel et al, 2005;

Werry & Gerhards, 1992; Werry, 1991). Des Weiteren scheint die Prognose bezüglich einer Rezidivfreiheit für an ERU erkrankte Augen besser zu sein, wenn leptospirale Antikörper aus dem entfernten Glaskörpermaterial nachgewiesen werden können. Aus diesem Grund wird eine präoperative Untersuchung von Glaskörpermaterial oder Kammerwasser mittels MAT vorgeschlagen, um für die Vitrektomie nur Augen mit nachweisbaren leptospiralen Antikörpern zu selektieren (Tömördy et al, 2010).

Ziel der vorliegenden Studie war es, zum einen den Einfluss unterschiedlicher Labore auf die Ergebnisse von MAT- und PCR-Untersuchungen zu evaluieren und zum anderen Langzeitergebnisse der Vitrektomie basierend auf ophthalmologischen Nachuntersuchungen zu erheben.

Die vorliegende Dissertation vergleicht im Kapitel I die Ergebnisse von PCR- und MAT-Untersuchungen unterschiedlicher Labore miteinander. Die Übereinstimmung der verschiedenen Labore in den einzelnen Nachweismethoden wird herausgestellt, wobei die Faktoren verdeutlicht werden, die diese Nachweismethoden beeinflussen. In Kapitel II werden ophthalmologische Langzeituntersuchungen von vitrektomierten Pferden präsentiert und der Einfluss von Geschlecht, Alter und Rasse der Pferde, die Bedeutung der MAT- und PCR-Ergebnisse und der präoperativen Vorschädigungen der vitrektomierten Augen evaluiert.

2 Literaturübersicht

2.1 Equine rezidivierende Uveitis

2.1.1 Definition

Die ERU ist eine entzündliche Augenerkrankung des Pferdes, die alle Uveastrukturen betreffen kann. Auf einen akuten, entzündlichen Schub folgt eine unterschiedlich lange, entzündungsfreie Periode, bevor eine erneute, akute Uveitis einsetzt (Lavach, 1990; Wollanke & Gerhards, 2009). Die Diagnose einer ERU wird somit nicht nur aufgrund gegenwärtiger klinischer Symptome gestellt, sondern ebenfalls auf Grundlage dokumentierter Krankheitsverläufe (Spiess, 2010).

2.1.2 Einteilung, Symptome und Verlauf

Die ERU wird in verschiedene klinische Unterformen eingeteilt. Eine Einteilung unterscheidet zwischen der „klassischen ERU“, der „schleichenden ERU“ und der „posterioren ERU“. Die „Klassische ERU“ ist hauptsächlich von akuten, sehr schmerzhaften Perioden okularer Entzündung geprägt, die hauptsächlich Iris, Ziliarkörper und Choroidea betreffen. Die „Schleichende ERU“ wird hingegen als eine milde, nahezu schmerzfreie Entzündung okularer Strukturen definiert. Letztgenannte Form kommt hauptsächlich beim Appaloosa und bei Kaltblupferden in den USA vor. Bei der „Posterioren ERU“ sind vorwiegend Glaskörper, Choroidea und Retina von entzündlichen Reaktionen betroffen. Warm- und Kaltblüter, sowie aus Europa in die USA importierte Pferde zeigen diese Form der ERU (Gilger, 2010).

Eine andere Einteilung orientiert sich strikter anhand betroffener Augenabschnitte und wird unterschieden in die „vordere Uveitis“ als eine sehr schmerzhafte Iritis, die „intermediäre Uveitis“ als eine schmerzlose Zyklitis und die selten vorkommende „hintere Uveitis“ als eine

Choroiditis (Wollanke & Gerhards, 2009). Andere Autoren unterscheiden neben der „vorderen Uveitis“ und der „intermediären Uveitis“ eine „Panuveitis“, die aufgrund einer Miosis zunächst als eine „vordere Uveitis“ eingeordnet wird (Spiess, 2010; Tömördy et al, 2010).

Weitere Symptome die in sehr großer Variabilität auftreten können sind eine Hyperämie der Konjunktiven, Trübungen der Hornhaut, der vorderen Augenkammer und der Linse, Hornhautvaskularisation, Miosis, Irisödem, Glaskörpertrübungen und Glaskörperverflüssigungen, Netzhautfalten bis hin zu einer Netzhautablösung (Cook et al, 1983; Gilger, 2010; Von Borstel et al, 2005; Von Borstel, 2010; Wollanke & Gerhards, 2009). Die ERU stellt somit eine der häufigsten Ursachen für einen Visusverlust beim Pferd dar (Lavach, 1990).

2.2 Ursachen

Bis heute werden die Ursachen der ERU kontrovers diskutiert, es kommen sowohl infektiöse als auch nicht-infektiöse Auslöser in Betracht. Neben genetischen Faktoren und autoimmunen Komponenten der Erkrankung ist vor allem die intraokulare Infektion mit Leptospiren eine vielfach erwähnte Ursache.

2.2.1 Genetik

Beobachtungen zeigen, dass einige Pferderassen häufiger an ERU erkranken als andere (Gilger, 2010). So besitzen Appaloosas ein 6,4 fach höheres Risikoverhältnis eine Uveitis zu entwickeln im Vergleich zu Vollblütern (Angelos et al, 1988). In einer anderen Studie zeigten Appaloosas sogar ein 8,3 fach höheres Risikoverhältnis eine Uveitis zu entwickeln im Vergleich zu allen anderen untersuchten Pferderassen (Dwyer et al, 1995a). Später konnten bei dieser Pferderasse drei genetische Marker mit dem Phänotyp der ERU in Verbindung gebracht werden, die einzelnt und kombiniert getestet bis zu 80 % der an ERU erkrankten Appaloosas und bis zu 75% der nicht an ERU erkrankten Appaloosas identifizierten (Fritz et al, 2014). Auch bei deutschen Warmblutpferden, die an ERU erkrankt waren, konnte ein signifikanter Einzelnukleotid-Polymorphismus auf dem equinen Chromosom 20 in direkter Nachbarschaft zu IL-17A und IL-17F festgestellt werden (Kulbrock et al, 2013a).

2.2.2 Intraokulare Leptospireninfektion

Eine der meist diskutierten Ursachen der ERU ist die intraokulare Leptospireninfektion. Die ersten Nachweise dieser Infektion beruhten auf erhöhten Antikörpertitern gegen Leptospiren im Serum von an ERU erkrankten Pferden im Vergleich zu augengesunden Pferden (Heusser, 1948; Rimpau, 1947). Zusätzlich konnte mit Hilfe des MAT in 43 bis 94 % der

Glaskörperproben von an ERU erkrankten Augen eine Erhöhung des Antikörpertiters gegen Leptospiren gemessen werden (Brandes et al, 2007; Brem et al, 1998; 1999; Gerding & Gilger, 2016; Gilger et al, 2008; Polle et al, 2014; Von Borstel et al, 2010; Wollanke et al, 1998; Wollanke et al, 2001). Durch die PCR war es möglich leptospirale DNS zu detektieren. In 40 bis 100 % der getesteten Glaskörper- bzw. Kammerwasserproben erkrankter Augen war ein Nachweis leptospiraler DNS möglich (Brandes et al, 2007; Faber et al, 2000; Polle et al, 2014; Von Borstel et al, 2010). Die kulturelle Anzucht vermehrungsfähiger Leptospiren aus Glaskörpermaterial oder Kammerwasser gelang in 21 bis 53 % der Proben von an ERU erkrankten Augen (Brandes et al, 2007; Brem et al, 1999; Faber et al, 2000; Hartskeerl et al, 2004; Wollanke et al, 2004; Wollanke et al, 2001).

Es gibt Hinweise darauf, dass intraokulare Leptospireninfektionen mit okularen Autoimmunreaktionen in Verbindung gebracht werden können. So konnte gezeigt werden, dass die leptospiralen Lipoproteine Lru A und Lru B aus an ERU erkrankten Augen eine Kreuzreaktivität mit Proteinen der Linse und der Netzhaut aufweisen (Verma et al, 2005; Verma et al, 2010).

2.2.3 Autoimmunreaktion

Verschiedene Autoren vermuten jedoch auch primäre autoimmune Reaktionen als Auslöser der spontanen ERU. So konnten in 72 % der Glaskörperproben von vitrektomierten ERU-Patienten Autoimmunreaktionen gegen das retinale S-Antigen und das retinale Interphotorezeptor Retinoid-Bindungsprotein nachgewiesen werden (Deeg et al, 2001). In einer darauffolgenden Studie konnte mit Hilfe des retinalen Interphotorezeptor Retinoid-Bindungsproteins eine Uveitis induziert werden (Deeg et al, 2002b). Und in späteren Untersuchungen wurden weitere

Autoantigene identifiziert, die Uveitiden bei Pferden und Ratten induzieren können (Deeg, 2009). Immunhistochemisch konnte in Augen, die an ERU erkrankt waren deutliche Hinweise auf immunmediierte Reaktionen gegen verschiedene okulare Autoantigene gefunden werden (Regan et al, 2012).

2.3 Therapie der ERU

Ziel der Therapie ist die Minimierung der okularen Entzündung, die Produktion und Freisetzung entzündlicher Mediatoren zu reduzieren und immunologische Mechanismen zu stoppen, um die Integrität der Blut-Augen-Schranke wiederherzustellen (Curling, 2011).

2.3.1 Konservative Therapie

Initial sollten lokal zykloplegische Mydriatika eingesetzt werden, um den Schmerz aufgrund von Krämpfen des Ziliarmuskels zu reduzieren und Synechien vorzubeugen (Schwink, 1992). Weiterhin werden bei intakter Kornea topische Kortikosteroide eingesetzt, um die Entzündung zu bekämpfen. Subkonjunktivale Injektionen mit Kortikosteroiden können sinnvoll sein, um eine möglichst hohe intraokulare Konzentration zu erreichen (Gerhards & Wollanke, 2001; Gilger & Michau, 2004; Rebhun, 1979). Systemisch werden Nichtsteroidale Antiphlogistika eingesetzt, bei besonders schweren Fällen der ERU kann auf systemische Kortikosteroide zurückgegriffen werden (Gilger & Michau, 2004).

Ob systemisch verabreichte antimikrobielle Substanzen eine intraokulare Konzentration erreichen, die eine bakterizide Wirkung nach sich zieht ist immer wieder kontrovers diskutiert worden. Intravenös verabreichtes Enrofloxacin scheint eine ausreichende Konzentration im

Serum und im Glaskörper von an ERU erkrankten Pferden zu erreichen, die über der minimalen Hemmkonzentration und der minimalen bakteriziden Konzentration von Leptospiren liegt. Einen signifikanten Einfluss auf die Anzüchtbarkeit von Leptospiren aus diesem Glaskörpermaterial scheint diese Enrofloxacinkonzentration jedoch nicht zu haben (Popp, 2013). Es wird davon berichtet, das einzelne intravitreale Injektionen von 4 mg Gentamicin Entzündungssymptome mildern und rezidivierende Schübe der ERU verhindern bzw. mildern können (Gilger, 2010; McMullen Jr, 2015).

2.3.2 Vitrektomie

Der operative Austausch von Glaskörpermaterial von an ERU erkrankten Augen mit gepufferter Kochsalzlösung, die Glutathion, Natriumbikarbonat und Glukose enthält, wurde zunächst durchgeführt um visuseinschränkende Glaskörpertrübungen zu entfernen und den rezidivierenden Charakter der Uveitis zu verringern (Werry, 1991). Nach den ersten positiven Erfahrungen konnte gezeigt werden, dass der Austausch des Glaskörpermaterials nicht nur Rezidive verringert, sondern über längere Zeiträume verhindern kann (Dorrego Keiter et al, 2017; Frühauf, 1998; Gerhards et al, 1999; Tömördy et al, 2010; Von Borstel et al, 2005; Werry & Gerhards, 1992; Winterberg & Gerhards, 1997).

Tömördy et al. (2010) untersuchten das Glaskörpermaterial von 47 vitrektomierten Augen mittels eines MAT auf Antikörper gegen Leptospiren und fanden in 40 Augen Antikörper. Insgesamt zeigten 33 der 40 positiv getesteten Augen keine erneuten Entzündungsschübe nach der Vitrektomie, wohingegen nur ein Auge der sieben negativ getesteten Augen keine erneuten Entzündungsschübe zeigten. Daraus wurde der Schluss gezogen, dass die Vitrektomie in Augen mit Antikörpern gegen Leptospiren erfolgreicher ist als in Augen ohne Antikkörper gegen

Leptospiren (Tömördy et al, 2010). Dorrego Keiter et al. zeigten jedoch, dass die Rezidivfreiheit nach einer Vitrektomie nicht im Zusammenhang mit einem positiven Nachweis von Antikörpern gegen Leptospiren steht (Dorrego Keiter et al, 2017).

2.3.3 Ciclosporin-Implantate

Ciclosporin A besitzt vor allem eine immunosuppressive Wirkung. Eine therapeutische Konzentration dieses Wirkstoffes kann am Auge jedoch nur nach intravenösen bzw. intravitrealen Injektionen oder nach Implantation eines Cyclosporin-Implantates erreicht werden (Gilger et al, 2000). Neben der Vitrektomie stellt die intravitreale oder suprachoroidale Implantation von Ciclosporin A freisetzenden Polyvinylalkohol- oder Silikonimplantaten eine alternative zur Reduktion uveitischer Schübe im Rahmen der ERU dar (Gilger & Michau, 2004). Durch das Ciclosporin-Implantat konnte in einem Großteil der Augen mit ERU die Sehfähigkeit aufrechterhalten werden (Gilger et al, 2010).

3 Material und Methode

3.1 Patientengut

3.1.1 Untersuchungsgut Kapitel I

In Kapitel I wurden die Glaskörperproben aus 93 an ERU erkrankten Augen von 78 Pferden in drei unterschiedlichen Laboren in Deutschland untersucht (siehe Seite 347, Pferdeheilkunde 32 (2016) 4 (July/August) 346-356).

3.1.2 Untersuchungsgut Kapitel II

In Kapitel II wurden die klinischen Daten von 235 Pferden ausgewertet, die zwischen 2005 und 2013 in der Klinik für Pferde der Tierärztlichen Hochschule Hannover vitrektomiert wurden (siehe Kapitel 5.3.1, Seite 19).

3.2 Vitrektomie

Die an ERU erkrankten Augen wurden einer „double port Pars plana Vitrektomie" unterzogen (siehe Kapitel 5.3.2, Seite 20).

3.3 Glaskörperprobengewinnung

Die Glaskörperproben wurden zu Beginn der Vitrektomie gewonnen, gleichmäßig in drei Proben aufgeteilt und an drei ausgewählte Labore versandt (siehe Seite 347, Pferdeheilkunde 32 (2016) 4 (July/August) 346-356).

3.4 Augenuntersuchung

Vitrektomierte Augen wurden mit einer Handspaltlampe, einem direkten Ophthalmoskop und einem Rückstoßtonometer untersucht (siehe Kapitel 5.3.4, Seite 22).

3.5 Nachweismethoden Leptospiren

3.5.1 MAT

Glaskörperproben wurden in zwei verschiedenen Laboren (L1 und L3) mit Hilfe des MAT auf Antikörper gegen Leptospiren untersucht (siehe Seite 348, Pferdeheilkunde 32 (2016) 4 (July/August) 346-356).

3.5.2 PCR

Die Labore L1 und L2 untersuchten die Glaskörperproben zusätzlich auf leptospirale Gene, so amplifizierte L1 das leptospirale lipL32-Gen und das L2 leptospirale gltA-Gen (siehe Seite 348, Pferdeheilkunde 32 (2016) 4 (July/August) 346-356).

3.5.3 Kultur

Das L3 führte eine kulturelle Anzucht potentiell vorhandener Leptospiren aus den Glaskörperproben durch (siehe Seite 348 Pferdeheilkunde 32 (2016) 4 (July/August) 346-356).

3.5.4 ELISA

Des Weiteren wurden die Glaskörperproben im L3 mittels ELISA auf leptospirale IgM, IgG und IgA Antikörper untersucht (siehe Seite 348, Pferdeheilkunde 32 (2016) 4 (July/August) 346-356).

3.6 Statistische Methoden

Die Ergebnisse der verschiedenen Labore wurden mit Hilfe der deskriptiven Statistik evaluiert und Übereinstimmungen der verschiedenen Labore anhand einer Kappa-Statistik untersucht (Landis & Koch, 1977) (siehe Seite 348, Pferdeheilkunde 32 (2016) 4 (July/August) 346-356).

Die Altersverteilung der Pferde wurde durch Binomialverteilung überprüft. Der exakte Fisher-Test und der Chi-Quadrat-Test wurden zur Evaluierung der Abhängigkeit verschiedener Eigenschaften auf das Auftreten von REC und NOIS benutzt. Zur Prüfung der Veränderungen zwischen prä- und postoperativem ERU-Score wurde der Rangsummentest von Wilcoxon verwendet. Schließlich wurde das Chancenverhältnis (Odds Ratio) nachweisbarer leptospiraler Antikörper und leptospiraler DNS für das Auftreten von REC und NOIS nach einer Vitrektomie berechnet (siehe Kapitel 5.3.5, Seite 24).

4 Kapitel I

Detection of intraocular leptospiral DNA, antibodies and Leptospira spp. in horses with equine recurrent uveitis in different laboratories

Eva I. A. Baake[1], Myriam von Borstel, Dr. med. vet.[1], Karl Rohn, Dr. med. vet.[2] and Bernhard Ohnesorge, Prof. Dr. med. vet., Dipl. ECEIM[1]

[1]Clinic for Horses, Universitiy of Veterinary Medicine Hannover, Foundation, Hannover, Germany

[2]Institute for Veterinary Biometry and Epidemiology, University of Veterinary Medicine Hannover, Foundation, Hannover, Germany

Veröffentlicht in: **Pferdeheilkunde, 32 (2016), 4, S. 346-345**

URL: http://www.hippiatrika.com/download.htm?id=20160407

DOI: 10.21836/PEM20160407

Anteil von Eva Baake an dieser Arbeit:

Eva Baake war beteiligt am Studiendesign und der Probengewinnung und -aufbereitung. Sie sammelte und analysierte die vorliegenden Daten, erstellte die Abbildungen bzw. Tabellen und schrieb das Manuskript.

4.1 Abstract

Equine recurrent uveitis (ERU) is an ocular disease characterized mainly by recurrent episodes of inflammation, alternating with quiescent episodes of various durations. There are numerous aetiological theories for the cause of ERU, but to date, the "classic ERU" is most commonly associated with an intraocular leptospiral infection. The pars plana vitrectomy has, so far, been the most promising treatment of ERU and is thought to be more successful in eyes with detectable leptospiral antibodies in the microscopic agglutination test (MAT) and therefore a preselection is recommended.

The objective of this study was to investigate whether different laboratories have an influence on the frequency of detection of leptospiral DNA and leptospiral antibodies from vitreal samples of horses with ERU and if these results are in accordance with cultural testing of the vitreal samples.

Vitreal samples were obtained from 93 eyes of 78 horses of various breeds and age with a clinical history of ERU. The vitreal samples were taken at the beginning of the vitrectomy to keep the dilution of the vitreal samples as low as possible. The samples were split and sent to three different laboratories in Germany (L1, L2 and L3). Laboratory 1 (L1) performed a polymerase chain reaction (PCR) and a MAT, laboratory 2 (L2) carried out a PCR and laboratory 3 (L3) performed a MAT, a cultural examination and an enzyme-linked immunosorbent assay on the vitreal specimens. The amount of dilution was evaluated by comparing the urea content of 34 vitreal samples with the urea content of the serum of each of the horses at the time of sampling. Furthermore, a serial dilution was performed with undiluted vitreal material of two enucleated eyes and sent to L1 for MAT and PCR.

The results of the PCR indicated a strong conformity between L1 and L2 (Kappa 0.95), both laboratories concurrently revealed 52 positive and 31 negative results and only two specimens showed deviating results. The MAT in L1 was positive in 51 % of the vitreal samples tested, whereas only 41 % of the vitreal samples were tested positive in L3. The conformity was, moreover, serovar- and breed-dependent. The Kappa was 0.67 when only including serovar Grippotyphosa, and was 0.82 for serovar Pomona (Tab. 5). Warmblood horses displayed a Kappa of 0.56 between L1 and L3, whereas the Kappa for the Icelandic horses was 0.91 between L1 and L3. The cultivation of leptospires was successful in 16 % of the vitreal specimens. The dilution of vitreal samples was measured to range between 1 and 2.1 with a standard deviation of 1.68.

Overall, 67 vitreal samples were tested simultaneously in all three laboratories. A positive result in at least one of the tests performed occurred in 57 % (38), whereas 43 % (29) of the vitreal specimens were tested negative in all five tests performed (PCR in L1 and L2, MAT in L1 and L3, and culture in L3).

The variation of agreement of the MAT between serovar Grippotyphosa and serovar Pomona and between Warmblood and Icelandic horses in L1 and L3 might be accounted for by the utilisation of congeneric, but different strains from the same leptospiral serotype. The dilution of vitreal specimens at the beginning of the vitrectomy might have influenced the results of this study. However, the serial dilution showed a negligible influence of a dilution as high as 1:3. The strong conformity of the results for PCR between L1 and L2 indicate a similar sensitivity of both PCR protocols.

Results indicated a strong inter-laboratory agreement when PCR was used for the detection of leptospiral DNA, whereas the MAT showed a variance of results that needs to be carefully regarded for interpretation. Therefore, basing a decision against a pars plana vitrectomy only upon a negative MAT result of anterior chamber fluid from a single laboratory could not be recommended.

5 Kapitel II

Long-term ophthalmologic examinations of eyes with equine recurrent uveitis after pars plana vitrectomy

Eva I. A. Baake[1], Myriam von Borstel, Dr. med. vet.[1], Karl Rohn, Dr. med. vet.[2], Michael H. Boevé, Prof. Dr. med. Vet., Dipl. ECVO[1] and Bernhard Ohnesorge, Prof. Dr. med. vet., Dipl. ECEIM[1]

[1]Clinic for Horses, Universitiy of Veterinary Medicine Hannover, Foundation, Hannover, Germany

[2]Institute for Veterinary Biometry and Epidemiology, University of Veterinary Medicine Hannover, Foundation, Hannover, Germany

Eingereicht in: **BMC Veterinary Research, zurzeit im Review**

Anteil von Eva Baake an diesem Manuskript:

Eva Baake war beteiligt am Studiendesign und führte die ophthalmologischen Untersuchungen durch. Sie sammelte und analysierte die Daten, erstellte die Abbildungen und schrieb das Manuskript.

5.1 Abstract

The objective of the study was to perform an ophthalmologic reexamination of eyes with a history of equine recurrent uveitis (ERU) after pars plana vitrectomy at the equine clinic of the University of Veterinary Medicine Hannover (TiHo) regarding the presence of vision and the presence of uveitis in those eyes after up to 123 months post vitrectomy.

Vitrectomy in eyes with ERU can influence the recurrent nature of the disease. Previous studies indicated success rates regarding the further absence of uveitis between 73.6 and 97.7 %. The present study aims at evaluating a long-term outcome for eyes that underwent a vitrectomy between 2005 and 2013 at the equine clinic of the TiHo. Aspects such as gender, age and breed of the horses, as well as leptospiral status and preexisting damage of operated eyes are described.

A total of 147 eyes were available for ophthalmologic reexaminations. After vitrectomy an improvement of the condition was seen in 133 eyes (91 %) and 14 eyes (9 %) showed at least one uveitic recurrence. A positive menace response and a positive dazzle reaction were observed in 128 of the 147 reexamined eyes (87 %). The number of episodes prior to the vitrectomy, the gender of the horse, and the leptospiral status of the eye showed no significant correlation with the prevalence of recurrences. Coat color, breed and age of the horse operated, as well as preexisting ocular alterations showed a significant influence on the chance of postoperative recurrences of uveitis. Horses with a coat color other than “chestnut” and “black & brown” had significantly more recurrences and the odds ratio for these horses to develop recurrences was also significantly higher.

The present study shows that a magnificent amount of the reexamined eyes remained visual (87 %) and a large proportion of the reexamined eyes (91 %) showed an improvement (73 % no further inflammatory symptoms, 18 % nonspecific ocular inflammatory symptoms) of the condition after vitrectomy. This study was also able to suggest that genetic factors, as well as age and preexisting ocular alterations, influence the long-term prognosis of eyes with ERU after vitrectomy.

5.2 Background

Equine recurrent uveitis (ERU) is an ocular disease in horses of different breeds with a prevalence between 2.7 and 7.6 % in Europe (Alexander, 1990; Szemes & Gerhards, 2000). Recurrent episodes of intraocular inflammation are painful, can lead to chronic damage to ocular structures and, hence, impair vision sooner or later (Angelos et al, 1988; Gerhards & Wollanke, 2001; Von Borstel et al, 2005; Wollanke & Gerhards, 2009). The impact of ERU on affected horses and their owners regarding the impairment of vision, the loss of use and the financial loss is significant (Gerding & Gilger, 2016). The immediate therapeutic target during an acute episode of ocular inflammation is to control the pain and keep the damage to intraocular structures as little as possible by immediately minimizing the ocular inflammation. This can be accomplished by the use of local mydriatic cycloplegics, topical corticosteroids and systemic nonsteroidal anti-inflammatory drugs. In addition to the acute therapy, only the prevention of future recurrences can effectively prevent further intraocular damage (Curling, 2011). Besides enucleation, vitrectomy and the implantation of a cyclosporine drug delivery device are currently the only therapeutic modalities that may prevent recurrences (Dorrego Keiter et al, 2017; Frühauf, 1998; Gilger & Allen, 1998; Gilger et al, 2010; Tömördy et al,

2010; Von Borstel et al, 2005; Werry & Gerhards, 1992; Werry, 1991; Winterberg & Gerhards, 1997).

Equine recurrent uveitis is categorized into three different clinical subtypes in the literature: The "classic ERU," the "posterior ERU" and the "insidious ERU." If horses show acute and painful episodes of intraocular inflammation affecting posterior and anterior parts of the eye, with recurrence after non-inflammatory episodes, ERU is considered to be the "classic ERU." If intraocular inflammation is primarily affecting the vitreous body, choroid and retina, it is considered to be "posterior ERU". This type of ERU is seen mostly in Warmbloods, draught breed horses and horses imported from Europe to the USA. "Insidious ERU" is characterized as a mild and almost painless intraocular inflammation, regularly seen in the Appaloosa and draught breed horses in the USA (Gilger, 2010).

Vitrectomy (single port, double port) in equine eyes suffering from ERU has been performed in Europe for more than two decades. During vitrectomy, the vitreous and the inflammatory debris it contains is mostly replaced by a buffered salt solution with glutathione, sodium bicarbonate and glucose[L], mixed with gentamicin[M] (Frühauf, 1998; Von Borstel et al, 2005; Werry & Gerhards, 1992; Werry, 1991). Initially, vitrectomy was expected to reduce and alleviate recurrences, but not to prevent them entirely (Werry, 1991). It was shown subsequently that vitrectomy was able to alter the recurrent nature of the disease in the majority of cases and, therefore, to prevent additional ocular pain and damage to intraocular structures (Frühauf, 1998; Werry & Gerhards, 1992; Winterberg & Gerhards, 1997). Besides preservation of vision, it has been implemented in the literature over the years that the vitrectomy is only deemed successful when no uveitic recurrences occur postoperatively (Tömördy et al, 2010; Von Borstel et al, 2005).

Since 1991, six studies have been published regarding the long-term prognosis after vitrectomy (Frühauf, 1998; Tömördy et al, 2010; Von Borstel et al, 2005; Werry & Gerhards, 1992; Winterberg & Gerhards, 1997). And no recurrences of uveitic symptoms were recognized in these studies in a range between 73.6 % (Tömördy et al, 2010) and 97.7 % (Winterberg & Gerhards, 1997) of the eyes evaluated. These variations may be due to different patient populations, operating methods, periods between the surgery and the follow-up, or variations in study designs. The aim of the current study was to provide a differentiated long-term prognosis based mainly upon ophthalmologic reexaminations and depending on gender, age and breed of each horse, the leptospiral status and preexisting alterations in eyes that underwent a double-port pars plana vitrectomy.

5.3 Materials and methods

A total of 284 eyes from 235 horses were vitrectomized between 2005 and 2013 at the equine clinic of the University of Veterinary Medicine of Hannover. Signalement and clinical data were collected from medical records.

5.3.1 Study population

All horses had a clinical history of at least two uveitic episodes of unknown etiology according to the referring veterinarian. A complete ophthalmologic examination was performed, including slit lamp biomicroscopy, direct ophthalmoscopy and tonometry in the clinic for horses prior to the vitrectomy to evaluate the extent of ocular changes and the current inflammatory status of the eye to determine the indication for vitrectomy. Vitrectomy was only performed in eyes without any signs of active intraocular inflammation.

Horses were stabled two days prior to the vitrectomy. Preoperative topical treatment consisted of ophthalmic ointment containing dexamethasone, neomycin and polymyxin B[A] eight times a day and atropine ophthalmic drops[B] (0.5 %, b.i.d.). Horses were held off feed for about six hours prior to the vitrectomy. Flunixin-meglumine[C] was administered intravenously in a dosage of 1.1 mg per kg body weight approximately one hour prior to the surgery. In addition, sulfadimethoxine and trimethoprim[D] was administered orally in a combined dosage of 20 mg per kg body weight approximately 3 hours prior to the surgery.

5.3.2 Surgical procedure

Horses were placed under general anesthesia in lateral recumbency with the affected eye facing up. The eyelashes were cut and the nasolacrimal duct and the conjunctival sac were flushed with a mild antibacterial solution containing sodium chloride, sodium mono- and dihydrogen phosphate and benzalkonium chloride[E], subsequently with a 1 % povidone-iodine solution and, finally, with saline solution mixed with gentamicin (4 mg/ml). The eye was surgically draped and an eye speculum was used to open the eyelids. A cunjunctival flap was prepared to expose the sclera 11 to 13 mm from the limbus at the area of the pars plana ciliaris. Two intrascleral cruciate sutures[I] were preplaced at the sclerotomy site at the 11 and 1 o'clock position to allow a fast closure after removal of instruments. A lancet-shaped blade was used to prepare the two transscleral ports to enable a double-port vitrectomy. The temporal sclerotomy was used as an irrigation port. The vitrectomy handpiece[J] of an oscillating vitrector was entered into the vitreous cavity through the nasal sclerotomy port. In order to control the vitrectomy procedure, an indirect binocular ophthalmoscope[K] was used to visualize the vitrectomy tip through the dilated pupil. All accessible vitreal material, especially uveitic effusions, was aspirated and replaced by buffered saline solution[L] mixed with gentamicin[M] (160 µg/ml). The vitreous cavity was flushed with 250 to 500 ml of the flushing solution, depending on the amount of vitreal material that needed to be removed. Both sclerotomies were closed immediately with the two preplaced intrascleral cruciate sutures after the removal of the vitrectomy handpiece and irrigation port. The conjunctiva was closed with a continuous suture[N] and 40 mg of gentamicin[M] and 4 mg of dexamethasone[O] were injected into the ventral subconjunctiva.

Vitrectomized eyes were reevaluated every day during hospitalization by slit lamp biomicroscopy and direct ophthalmoscopy. In straightforward cases topical therapy with ophthalmic ointment containing dexamethasone, neomycin and polymyxin B[A] was conducted q.3.h. for five days postoperatively, q.6.h. for another seven days and t.d.s. for another 14 days subsequently at the home stable. Oral flunixin-meglumine[C] therapy was continued until one day postoperatively at the full dosage (1.1 mg/kg body weight b.i.d.) and the second and third day postoperatively at the half dosage (0.55 mg/kg body weight b.i.d.). Sulfadimethoxine and trimethoprim[D] were administered for four days (20 mg/kg b.i.d.).

Vitreal samples were obtained during the first few seconds after the start of the vitrectomy via the specimen collection vial of the vitrectomy system. Vitreous material was immediately sent to an external laboratory[P] to perform a microscopic agglutination test (MAT) test for intraocular leptospiral antibodies and a polymerase chain reaction (PCR) detecting leptospiral DNA. A total of 254 vitreous samples were available for MAT and PCR.

5.3.3 Contacting for re-examination

For the current study owners were contacted by telephone and interviewed with the aid of a standardized questionnaire. They were asked about the postoperative clinical course of the vitrectomized eye/eyes, whether any signs of ocular inflammation or other ocular problems had been noticed and/or may have been diagnosed by a veterinarian. Three different options were used to differentiate between possible outcomes: free of complaints (FOC), recurrences (REC) or non-specific ocular inflammatory symptoms such as conjunctivitis or keratitis (NOIS) (Tab. 1). After answering the questionnaire, the owners were offered an ophthalmologic check-up of their horse at their home stable.

Tab. 1: Possible postoperative outcomes after the vitrectomy.

postoperative progression of ERU	abbreviation	
signs of ocular inflammation, similar to the episodes prior to the vitrectomy	REC (recurrence of uveitis)	
mild signs of ocular inflammation, not comparable to the episodes prior to the vitrectomy (e.g. epiphora, chemosis, due to conjunctivitis, keratitis)	NOIS (non-specific ocular inflammatory symptoms)	IMP (improvement)
no signs of ocular inflammation	FOC (free of complaints)	IMP (improvement)

5.3.4 Ophthalmologic examination

A complete ophthalmologic examination was performed to evaluate the vitrectomized eyes for the current study, including slit lamp biomicroscopy, direct ophthalmoscopy, tonometry and induction of mydriasis using tropicamide^{R}. Findings were documented and an ERU-score established previously was used to estimate the status quo of each eye. The score was applied for different ocular compartments (iris, lens, vitreous and fundus oculi) and for other chronic changes and resulted in a total-score for each eye. Values ranged from zero (no abnormality detected) to five (high grade abnormalities) (Tab. 2) (Kulbrock et al, 2013b; Von Borstel et al, 2010). This score-system was also retrospectively applied to get a preoperative score for each eye based on documented examinations previously to the vitrectomy.

Tab. 2: Score for chronic uveitis findings of the most affected ocular compartements (Score by M. von Borstel (Kulbrock et al, 2013b; Von Borstel et al, 2010)).

score	iris	lens	vitreous	fundus oculi	other chronic changes
0	no abnormality detected	no abnormality detected	no abnormality detected	no abnormality detected	no abnormality detected
1	low-grade depigmentation; focal small synechia	focal capsular cataract; focal iris residues	low-grade liquefication with several floating filamentous strands of cell debris and inflammatory products	single focal chorioretinopathy („bullet hole lesion")	low-grade reduction in size of the bulb/ reduction in size of the anterior chamber
2	low-grade depigmentation; focal large-scale synechia	multifocal capsular/ subcapsular cataract; local bullous subcapsular/ cortical cataract; multifocal iris residues	low-grade liquefication; several curtain-like strands of cell debris and inflammatory products	a few focal chorioretinopathies ("bullet hole lesion"); small-area peripapillary chorioretinopathy ("butterfly lesion")	moderate reduction in size of the bulb/ reduction in size of the anterior chamber; local corneal haze
3	moderate depigmentation; several focal synechiae	local reticular capsular/ subcapsular cataract; local immature corticale/ nuclear cataract	moderate liquefication; moderate curtain-loke strands of cell debris and inflammatory products; low grade diffuse yellow haze	multiple focal chorioretinopathies; large-scale peripapillary chorioretinopathy	low-grade bulb enlargement of the anterior; laminar corneal haze; low-grade increase in intraocular pressure
4	moderate depigmentation; atrophy; circular synechiae	diffuse immature capsular/ subcapsular/ cortical and/ or nuclear cataract; lens subluxation	high-grade liquefication; moderate curtain-like strands of cell debris and inflammatory products; moderate diffuse yellow haze	laminar chorioretinopathies/ laminar degeneration of the retina; peripapillary wrinkles of the retina; partial retinal detachment	moderate bulb enlargement/ enlargement of the anterior chamber; laminar coreal haze/ haab`s sriae; moderate increase in intraocular pressure
5	high-grade depigmentation; seclusio pupillae	mature / hypermature cataract; lens luxation	high-grade liquefication; high-grade visible strands of cell debris and inflammatory products; high-grade diffuse yellow haze	complete retinal detachment	high-grade bulb enlargement/ enlargemenet of the anterior chamber; laminar corneal haze/ haab`s striae; high-grade increase in intraocular pressure; phtisis bulbi

5.3.5 Statistical evaluation

Statistical analysis was performed with the SAS Enterprise Guide 7.1[S]. The binomial distribution was used to assess the age distribution among the horses vitrectomized. Fisher's exact test was used to evaluate the frequency of distribution in the group of horses vitrectomized between 2005 and 2013 and the patient population in the clinic for horses during the same period. Since not all vitrectomized eyes were available for follow-up information, two distinct groups were created to further specify and verify information. One group consisted of all horses with follow up information (A) and one only of the horses with ophthalmologic reexaminations (B). Fisher's exact test and the chi square test were used to evaluate the dependence of various characteristics on the frequency of REC and NOIS. The group NOIS was tested combined with REC against FOC and it was tested combined with FOC against REC. Wilcoxon signed rank test was performed to evaluate the trends between the height of the pre- and postoperative ERU-scores. Odds ratios were used to estimate the influence of detectable leptospiral antibodies and detectable leptospiral DNA on the appearance of REC and NOIS.

5.4 Results

5.4.1 Patient population

The retrospective patient population was composed of various breeds and comprised 102 mares (124 eyes), 109 geldings (132 eyes) and 24 stallions (28 eyes). The major proportion of the vitrectomized horses were colored "black & brown" (57 %) (Tab. 3). Most horses were between four and seven years old (106/235) at the vitrectomy of the first eye (Tab.4).

Tab. 3: Gender, breed and coat color distribution of the vitrectomized horses and the patient population of the clinic for horses between 2005 and 2013. *p<0.05, **p<0.01, ***p<0.001 (fisher's exact test); + brown, black-brown, black; ++gray, dappled, others, undocumented

	vitrectomized horses			patient population, clinic for horses universitiy of veterinary medicine hannover
gender	n	(%)		(n / %)
mares	102	(43)	** (mares vs. geldings)	(6827 / 53)
geldings	109	(46)	*** (geldings vs. stallions)	(4341 / 34)
stallions	24	(10)		(1749 / 14)
female	102	(43)	**	(6827 / 53)
male	133	(56)		(6090 / 47)
breed	n	(%)		(n / %)
warmbloods	123	(52.3)	*** (vs. icelandic horses); * (vs. frisian horses); *** (vs. trotters); * (vs. pura raza espanola); ** (vs. knabstruppers)	(7941 / 62.1)
icelandic horses	32	(13.6)		(584 / 4.6)
ponies	29	(12.3)		(1492 / 11.7)
frisian horses	10	(4.3)		(325 / 2.5)
trotters	9	(3.8)		(138 / 1.1)
pura raza espanola	7	(3)		(154 / 1.2)
quarter horse / paint	6	(2.6)		(408 / 3.2)
arabic horses	5	(2.1)		(457 / 3.6)
knabstruppers	3	(1.3)		(24 / 0.2)
thoroughbred	3	(1.3)		(503 / 3.9)
pinto	1	(0.4)		(50 / 0.4)
others	7	(3)		(711 / 5.5)
coat color	n	(%)		(n / %)
black & brown+	135	(57)	* (vs. others++)	(5513 / 50)
others++	53	(23)		(3077 / 28)
chestnut	47	(20)		(2527 / 22)

Tab. 4: Age distribution of the vitrectomized horses.

age group	age in years	n	(%)
1	0-3	27	(11,5 %)
2	**4-7**	**106**	**(45,1 %)**
3	8-11	47	(20 %)
4	12-15	33	(14 %)
5	16-19	15	(6,4 %)
6	20-23	6	(2,6 %)
7	24-27	1	(0,4 %)
8	≥28	0	(0 %)

Horses with an age between four and seven years old were significantly overrepresented compared to every other age group ($p = < 0.0001$) (Tab. 4). Compared to the patient population in the clinic for horses between 2005 and 2013, vitrectomy was significantly more often performed in male than in female horses (males: $p = 0.0037$; geldings: $p = 0.0049$; stallions: $p = 0.0001$) and significantly more often in Icelandic horses ($p = < 0.0001$), Frisian horses ($p = 0.0411$), Trotters ($p = 0.0004$), Pura Raza Espanola ($p = 0.0118$) and Knabstruppers ($p = 0.0061$) than in Warmbloods (Tab. 3). More than half of the horses vitrectomized were Warmbloods (52.3 %, 123/235). The most frequent coat color among the horses vitrectomized in this study was “black & brown” and these were significantly more often vitrectomized than horses with the coat color “other” ($p = 0.031$), when compared to the patient population of the equine clinic in the same time period (Tab. 3).

5.4.2 Follow-up

Follow-up information of 201 horses (246 eyes vitrectomized) was available either through contacting the owner, through later hospitalizations and / or an ophthalmologic check-up at the home stable was performed as described on 116 horses (147 eyes) (Fig. 1).

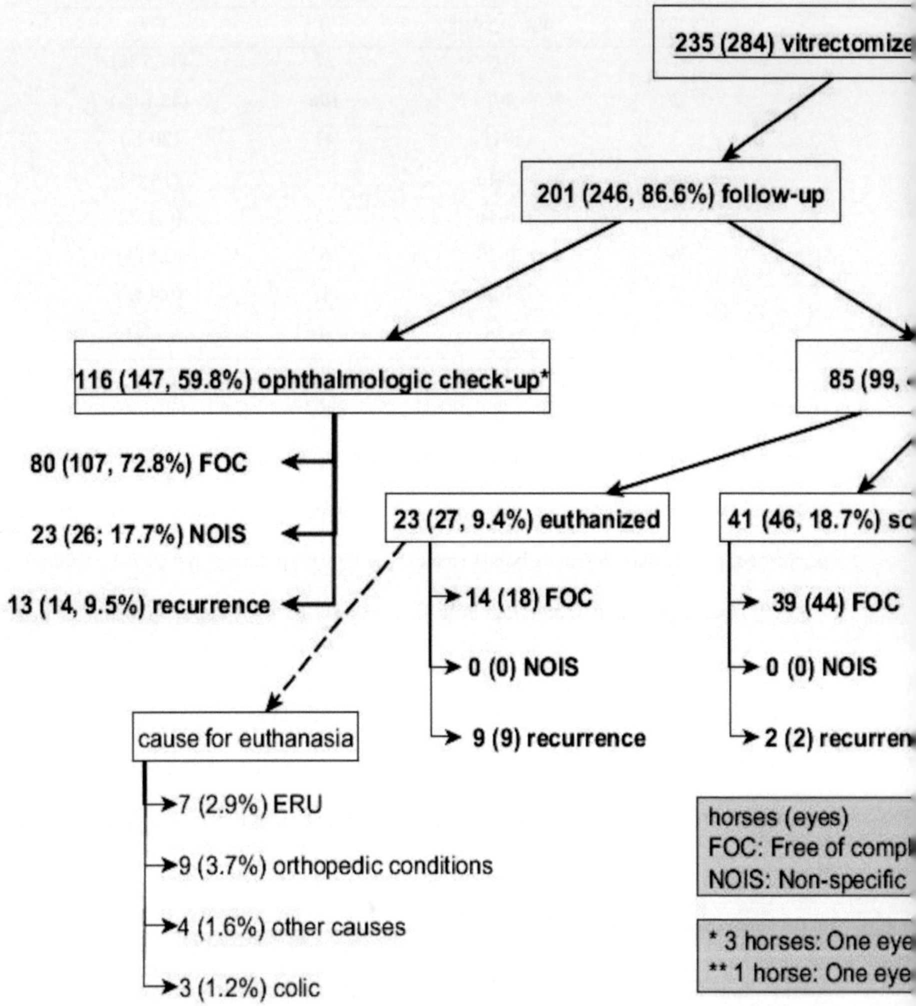

Fig. 1: Overview about the horses and eyes available for follow-up anc

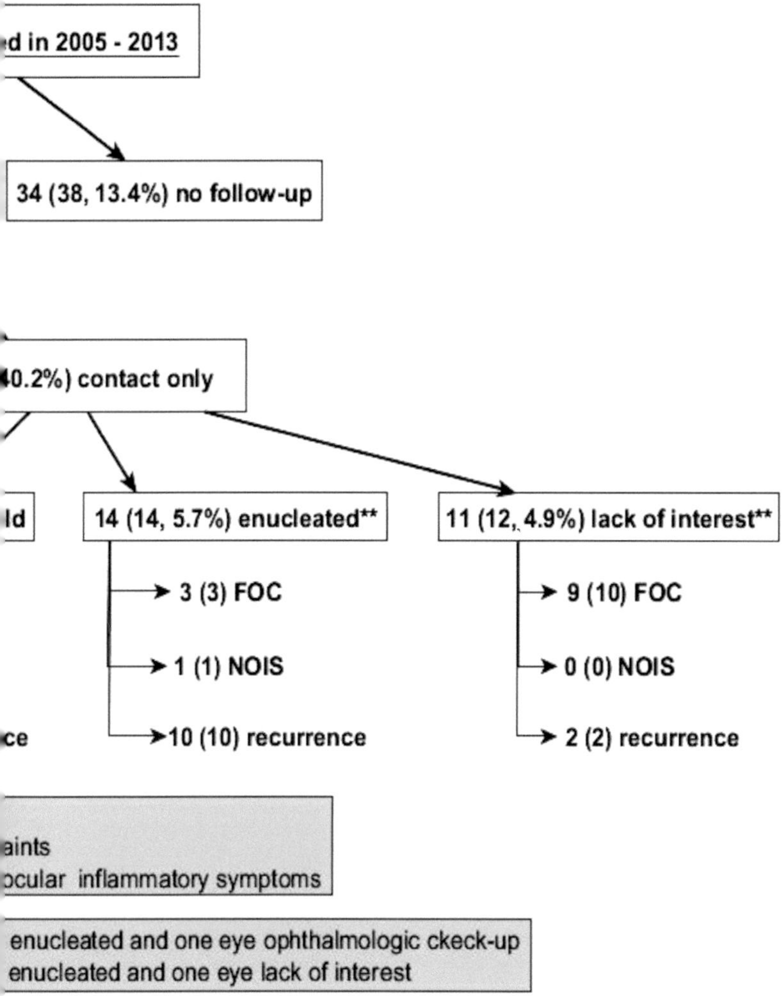

for ophthalmologic check-up.

The median time between vitrectomy and contact was 64 months (minimum 17 months, maximum 123 months). Looking only at the reexamined eyes, a total of 73 % (107/147) of the eyes showed no signs of uveitis or other ocular inflammation/problems at all (FOC). Mild signs of ocular inflammation not comparable to the episodes prior to the vitrectomy such as keratitis or conjunctivitis (NOIS) were shown by 18 % (26/147) of the eyes. So an improvement of the condition was achieved in 91 % (133/147) of the eyes (IMP), whereas 9 % (14/147) of the eyes showed signs of ocular inflammation similar to the episodes prior to the vitrectomy (REC). Taking all eyes into account where information was available, 74 % (182/246) were FOC and 11 % (27/246) of the eyes were categorized as NOIS, therefore 85 % (209/246) of the eyes were categorized as IMP, while REC was shown in 15 % (37/246) of the eyes. No significant differences between the follow-up group (A) and the ophthalmologic check-up group (B) could be shown (Fig. 2).

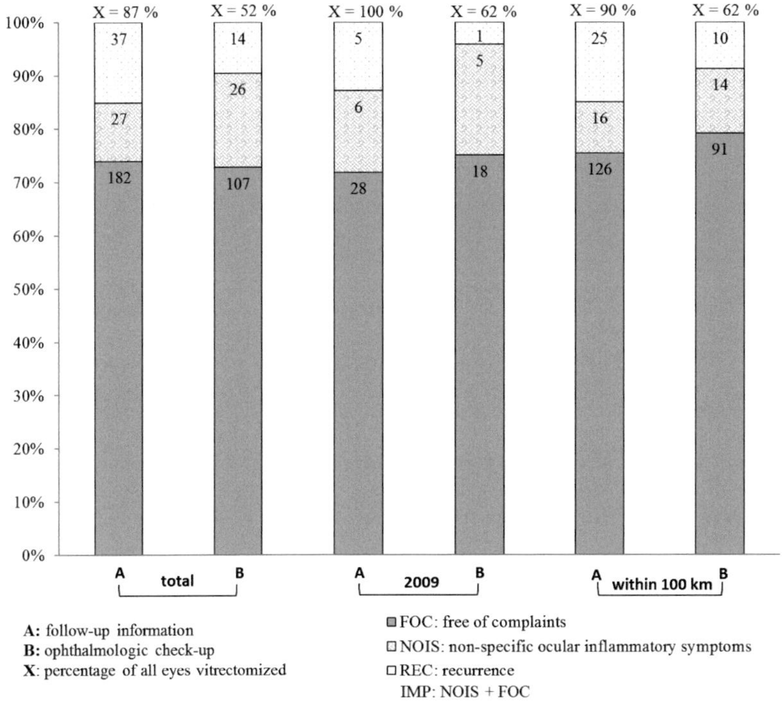

Fig. 2: Number of eyes with "FOC", "NOIS" and "REC", stratified by year and two different radii around the clinic.

5.4.3 Stratification

Because not all vitrectomized horses were available for follow-up information, the available information was stratified for different factors that were not able to affect the outcome of the vitrectomized eye. This was done to verify the success-rates concerning the absence of uveitic symptoms after the vitrectomy.

Since 100 % of the horses vitrectomized from the year 2009 (n = 39) were available for follow-up information, this group of eyes was used for a stratification (Fig. 2).

In addition, the owners place of residence was used to further stratify the follow-up information. A radius of 100 km (90 % follow-up-rate) around the clinic was used. No significant differences could be shown between both stratifications (total vs. 2009 and within 100 km) and also between the ophthalmologic check-up group (B) and the follow-up and ophthalmologic check-up group (A).

5.4.4 ERU-Score

All reexamined eyes were scored with the ERU-score (Tab. 1). This post-vitrectomy-score was compared to the pre-vitrectomy-score, which was based on the documented examinations prior to the vitrectomy. The Wilcoxon signed rank test showed no significant differences between the total-score from pre- (2.8 ± 1.1) to post-vitrectomy (2.9 ± 1.2) for the reexamined eyes (B). The lens-score increased significantly from pre- (1.5 ± 1.1) to post-vitrectomy (2.8 ± 1.3), whereas the vitreous-score decreased significantly from pre- (2.5 ± 1.3) to post-vitrectomy (1.7 ± 1.7). All other compartments did not show any significant changes from pre- to post-vitrectomy.

5.4.5 Sight

A negative dazzle reaction combined with a negative menace response was observed in 4 % (6/147) and a positive dazzle reaction combined with a negative menace response was observed in 9 % (13/147) of the eyes reexamined (B). One of the six eyes with a negative dazzle reaction and a negative menace response had a fundus-score of 5, which was defined as a complete retinal detachment and the other five horses had a lens-score of 5, which was defined as a mature or hypermature cataract or a lens luxation (Tab. 1).

The group of eyes with a positive dazzle reaction combined with a negative menace response had four eyes with a lens-score of 4 (diffuse immature capsular / subcapsular / cortical and / or nuclear cataract; lens subluxation), eight eyes with a lens-score of 5 and one eye with a fundus-score of 4 (laminar chorioretinopathies / laminar degeneration of the retina; peripapillary wrinkles of the retina; partial retinal detachment).

5.4.6 Number of uveitis episodes

The number of uveitis episodes pre-vitrectomy did not influence the occurrence of REC significantly.

5.4.7 Gender, breed, coat color, age

The gender did not influence the rate of REC significantly ($p > 0.05$). Overall, 17 % of the female eyes showed REC (18/103) and 83 % of the female eyes improved after the vitrectomy (IMP: 85/103; FOC: 68/103 + NOIS: 17/103). The number of REC was 13 % (19/143) in vitrectomized eyes of male horses and 87 % of the eyes of male horses improved after the vitrectomy (IMP: 124/143; FOC: 114/143 + NOIS: 10/143).

All Knabstruppers (3/3) in the present study showed REC and where therefore excluded from the further statistics.

Significantly more REC were noticed in the eyes of horses with the coat color "other" (12/55, 22 %) compared to eyes of horses with a "black & brown" coat color (15/141, 11 %) ($p = 0.0264$).

Older horses tended to have more REC than younger horses. Horses older than 20 years showed a significantly higher number of eyes with REC in contrast to horses aged four to seven years ($p = 0.0477$) (Fig. 3).

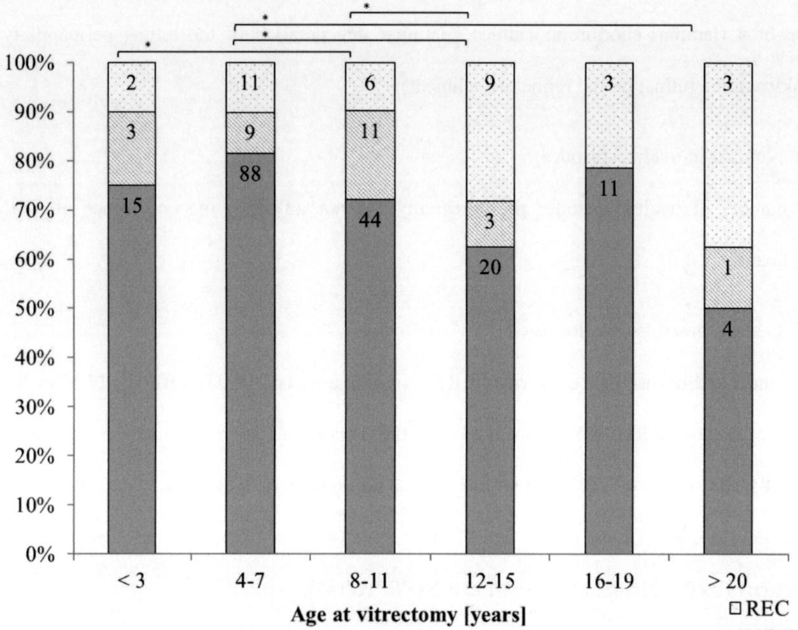

Fig. 3: Number of eyes with "FOC", "NOIS" and "REC" by age at vitrectomy.
*p<0.05, **p<0.01, ***p<0.001 (fisher's exact test)

5.4.8 ERU-score and time-point of recurrence

The higher the pre-operative ERU-score, the higher the rate of REC was. Eyes with a total score of three and five had significantly more REC than horses with a score of two (p = 0.0326, p = 0.0221, respectively) (Fig. 4).

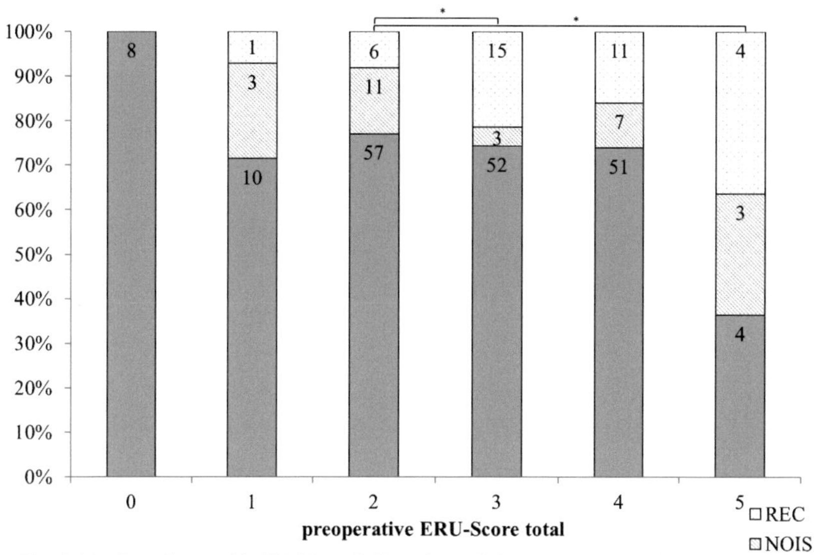

Fig. 4: Number of eyes with "FOC", "NOIS" and "REC" by ERU-Score at vitrectomy. *p<0.05, **p<0.01, ***p<0.001 (fisher's exact test)

The mean time between vitrectomy and the first REC was 12.8 (± 22.8) months. The latest onset of REC was 95 months after the vitrectomy; this was observed in two eyes. A total of 17 REC (6.9 %) were noticed during the first two months. Six REC occurred between the third and the sixth month post-vitrectomy. The Kaplan-Meier value showed that the probability of a non-recurrence of intraocular inflammation after vitrectomy decreased over time. The highest decrease happened during the first two months. After 20 months, the probability seemed to reach a plateau, which dropped down again after approximately 50 months. After 123 months the probability of a non-recurrence was 80.79 % (Fig. 5).

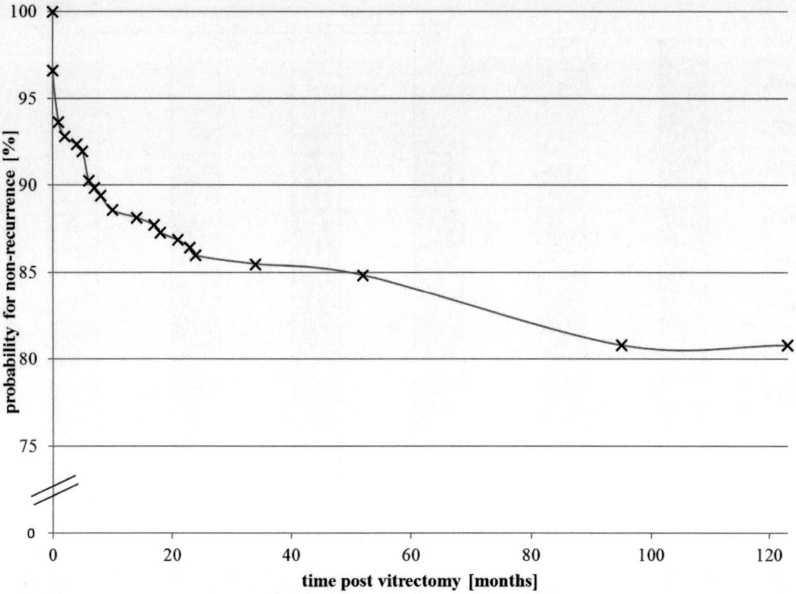

Fig. 5: Kaplan-Meier estimation of the probability for non-recurrence.

5.4.9 Leptospiral-status

Horses younger than three years were significantly more often leptospiral-positive than horses aged 12 to 15 years ($p = 0.0176$) and older than 20 years ($p = 0.0014$) (Fig. 6). The color distribution showed no significant differences between the age groups.

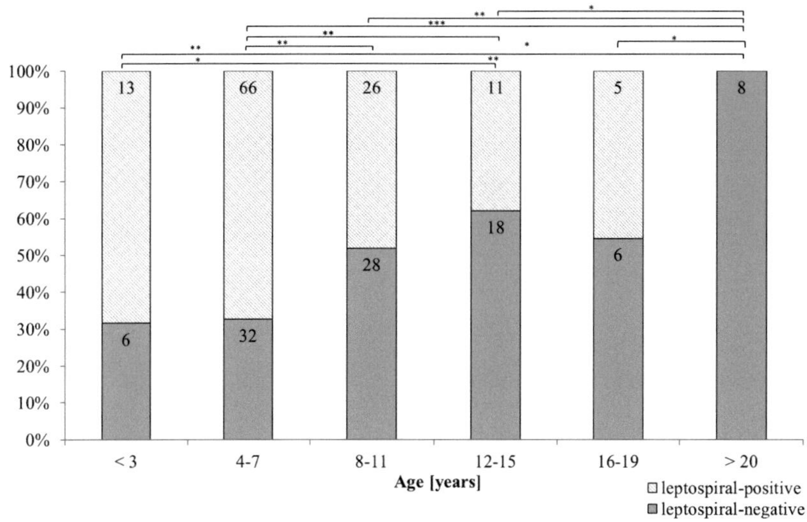

Fig. 6: Distribution of leptospiral-positive and leptospiral-negative eyes between different age groups. *p<0.05, **p<0.01, ***p<0.001 (fisher's exact test)

Leptospiral-negative eyes (16/98, 16 %) did not show significantly more recurrences than leptospiral-positive eyes (13/121, 11 %).

The odds ratio of developing REC after a vitrectomy was higher (not significantly) for leptospiral-negative eyes (OR: 1.925, CI: 0.8988, 4.1228). Horses where a serovar other than serovar Grippotyphosa was detected had a higher odds ratio to develop REC than horses where serovar Grippotyphosa was detected (OR: 2.1667, CI: 0.6065, 7.7407). The odds ratio for developing REC was also higher for Warmbloods when the eyes were leptospiral-negative (OR: 2.087, CI: 0.6945, 6.2713). Icelandic horses had a higher odds ratio to develop REC than Warmbloods (OR: 1.4637, CI: 0.5289, 4.0504). None of the latter odds ratios were significant.

The odds ratio was significantly higher to develop REC for the coat color “other” than for the coat colors “chestnut” or “black & brown” (OR: 2.6321, CI: 1.2596, 5.5003) (Fig. 7).

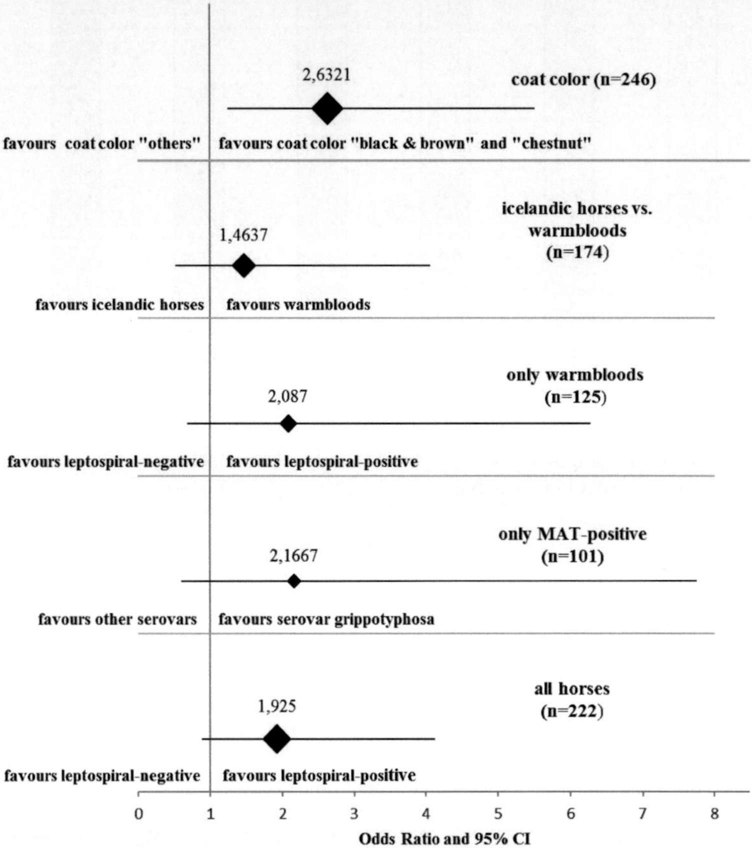

Fig. 7: Forest diagram showing the odds ratios for having a "REC" between different stratifications.

The outcome of the vitrectomy regarding the absence of recurrent episodes (REC) was tested for Warmbloods and for Icelandic horses depending upon the leptospiral status of the vitrectomized eyes.

The rate of IMP (improvement) was 88 % (121/137) for all Warmblood horses and 84 % (31/37) for all Icelandic horses. When only leptospiral-positive eyes were included, the rate of IMP was lower (79 %) for Icelandic horses (15/19) than for Warmblood horses (64/70; 91 %). This difference was not significant (p = 0.999).

5.5 Discussion

Follow-up surveys are the only way to provide a prognosis for horses with ERU that are going to have a vitrectomy. It is in the nature of follow-up surveys that not all subjects are available for follow-up information, and even fewer are available for follow-up examinations. Therefore, the rates for non-recurrence of uveitic symptoms in literature ranged between 73.6 and 97.7 % (Tab. 5) (Frühauf, 1998; Tömördy et al, 2010; Von Borstel et al, 2005; Werry & Gerhards, 1992; Winterberg & Gerhards, 1997). These rates were based on ophthalmologic reexaminations in only three of six studies, with the other studies not indicating how many horses were reexamined as part of the study design (Dorrego Keiter et al, 2017; Frühauf, 1998; Winterberg & Gerhards, 1997). The majority of information in the latter studies was gained through interviewing the owners by telephone (Tömördy et al, 2010; Von Borstel et al, 2005). The number of eyes evaluated in all of these studies was between ten and 134, and the period between the vitrectomy and the follow-up ranged between two and 90 months. Only one study gave a median time of 29 months of the evaluation after the vitrectomy (Dorrego Keiter et al, 2017). No recurrences of uveitic symptoms were recognized in these studies in a range between

73.6 % (Tömördy et al, 2010) and 97.7 % (Winterberg & Gerhards, 1997) of the eyes evaluated. These variations may be due to different patient populations, operating methods, periods between the surgery and the follow-up, or variations in study designs.

Tab. 5: Published studies regarding the long-term prognosis of horses with ERU after a vitrectomy.

	evaluated eyes	time post vitrectomy [months]	opthalmological reexaminations	no uveitic symptoms	conservation of visual acuity (not blind)
Werry et al. 1992	90.9 % (10/11)	not specified	not specified	100 %	no information
Winterberg et al. 1997	53.1 % (43/81)	6 to 67	100 %	97.7 %	72 % (31/43)
Frühauf et al. 1998	89.5 % (34/38)	5 to 60	79 %	85 %	85 % (29/34)
Borstel et al. 2005	89.3 % (50/56)	2 to 3	not specified	94 %	80% (40/50)
Tömördy et al. 2010	89.8 % (53/59)	3 to 90	not specified	73.6 %	no information
Dorrego-Keiter et al. 2017	100 % (134/134)	8 to 54 median 29	not specified	76.3 % (FOC + NOIS)	no information
Present study, Baake et al. 2017	86.6 % (246/289)	17 to 123 median 64	59.8 % (147/246)	**91 % IMP** 73 % FOC 18 % NOIS	87 % (128/147)

The aim of the current study was to obtain information of, and to perform a reexamination on as many eyes as possible after a period of at least 17 and up to 123 months post-vitrectomy. Follow-up information for 86.6 % (246/284) of all eyes vitrectomized between 2005 and 2013 was available. A total of 59.8 % (147/246) of these eyes were available for ophthalmologic reexaminations and 40.2 % (99/246) were contacted only via telephone (Fig. 1).

5.5.1 Ophthalmologic check-up group (B) vs. follow-up information group (A)

The number of eyes with REC was smaller in the ophthalmological check-up group (B: 9 % REC) compared to the follow-up information group (A: 15 % REC). This may partly be explained by the numbers of enucleation and euthanasia in the "information only" group (Fig. 2). Since any episode of uveitis may result in globe destruction and loss of vision and since uveitis in equines is generally highly painful, euthanasia may have been a reasonable choice for owners in these cases (Gerding & Gilger, 2016).

Since the ophthalmological reexamination for the current study was performed only once, it could not be clarified at what time after the vitrectomy the eyes showed recurrences. This might only be clarified by repetitive ocular examinations at regular intervals.

Whether NOIS is accounted for in the group of REC or the group of FOC could not be finally clarified. An argument that NOIS needs to be numbered among FOC is that inflammatory ocular signs of eyes with NOIS were certainly distinguished by owners from the episodes prior to the vitrectomy and that veterinarians diagnosed in the eyes of their horses another ocular problem besides uveitis (e.g. keratitis, conjunctivitis). Owners of a vitrectomized horse were presumably very sensitively attuned to recognizing signs of ocular inflammation in their horse. This aspect may provide a bias in the "information only" group. Anyway, a certain expectation of the owners regarding the success of the vitrectomy should not be underestimated. The FOC and NOIS (IMP) together represented 85 % (209/246) of the eyes vitrectomized (Fig. 2). With regard to the studies published about the long-term prognosis after vitrectomy, the rate for non-recurrence as FOC only ranged between 73 and 74 % and confirmed the results of Tömördy et al. (2010) who found a rate of non-recurrence of 73.6 % and of Dorrego-Keiter et al. (2017) with a rate of non-recurrence of 77.6 % (Dorrego Keiter et al, 2017; Tömördy et al, 2010).

Dorrego-Keiter et al. (2017) clearly stated that they include eyes with ocular problems other than uveitis (e.g. keratitis, conjunctivitis) as a non-recurrence (Dorrego Keiter et al, 2017). If NOIS is counted among FOC, the rate for non-recurrence ranged between 85 to 91 % in the present study, which is almost similar to the findings of Winterberg et al. (1997) Frühauf et al. (1998), and von Borstel et al. (2005). Nevertheless, some results of previous publications significantly alter from each other. This may not only be explained by different time periods between the vitrectomy and the ophthalmologic follow-up examination for these studies, but also by the method of evaluation (ophthalmologic examination vs. contact only). Tömördy et al. (2010) reexamined, respectively, contacted owners and referring veterinarians of the horses between 3 and 90 months post-vitrectomy, whereas Frühauf et al. (1998) examined 27 of 38 eyes of the horses vitrectomized between 5 and 60 months post-vitrectomy and von Borstel et al. (2005) reexamined or contacted owners after two to three months post-vitrectomy. Frühauf et al. (1998) reported that most of the REC were recognized during the first and second month post-vitrectomy. The present study confirmed that most of the REC occurred during the first two months post-vitrectomy (17 out of 37 eyes with REC) (Fig. 5). In addition, we also found that late REC occur, as a total of 20 eyes with REC were noticed later than the second month post-vitrectomy in the present study. Two eyes even showed REC after 95 months, but these were the only REC that were recognized later than 60 months post-vitrectomy. Assumed that the ocular examinations in the current study would have been performed only three months after each vitrectomy, the success rate (IMP) would have increased up to 93 %. Von Borstel et al. (2005) evaluated vitrectomized eyes only two to three months post-vitrectomy and, therefore, probably demonstrated a higher success rate (94 %) than in the present study. Since Tömördy et al. (2010) evaluated eyes up to 90 months post-vitrectomy, their rate for non-recurrence was less than presented by von Borstel et al. (2005). The longer the postsurgical period reviewed, and the more reexaminations performed, the lower the found rates for postsurgical non-recurrence are. The Kaplan-Meier values provide the impression of a rapid decrease of REC during the first six months post-vitrectomy and a slower, but steady decrease of the probability of non-recurrence from 89 % down to 80 % during the time between seven to 123 months (Fig. 5). Nevertheless surgeons, surgical methods, patients and owners' compliance may have influenced the rate for non-recurrence as well.

A limitation of the present study was the fact that not all horses were available for an ophthalmological reexamination. Therefore, the actual rate of REC might be higher or lower than determined. In addition, the memories of the owners may have been insufficient regarding the health issues of each horse. In order to minimize these limitations and to determine these influences, the available information was stratified by different factors (total, year 2009, radius of 100 km around the clinic). But regardless if eyes were reexamined or information was only based on phone calls, the decision concerning the possible postoperative outcome of an eye was in the end always based upon the owners statements.

No significant difference could be shown between the total number of REC and the two stratifications (100 km and the year 2009) (Fig. 2). Therefore, the assumption regarding the long-term prognosis in the current study was thought to be of high value.

5.5.2 ERU-Subtypes and scoring

In the literature, equine recurrent uveitis is categorized into three different clinical subtypes in literature: the "classic ERU," the "posterior ERU" and the "insidious ERU." If horses show acute and painful episodes of intraocular inflammation affecting posterior and anterior parts of the eye, with recurrence after non-inflammatory episodes, ERU is considered to be the "classic ERU." If intraocular inflammation is primarily affecting the vitreous body, choroid and retina, it is considered to be "posterior ERU". This type of ERU is seen mostly in Warmbloods, draught breed horses and horses imported from Europe to the USA. "Insidious ERU" is characterized as a mild and almost painless intraocular inflammation, regularly seen in the Appaloosa and draught breed horses in the USA (Gilger, 2010). In the current study, the major proportion of the horses belonged to the subtype "classic ERU" or "posterior ERU". It can be assumed that the 3 Knabstruppers belonged to the subtype "insidious ERU", these horses were therefore excluded from the further statistics.

The total ERU-score of all eyes examined from pre- to post-vitrectomy did not change significantly from pre- to post-vitrectomy in the current study, including the eyes with REC and NOIS. Since ophthalmologic changes tend to worsen in eyes with ERU during each inflammatory episode, if they are not vitrectomized, this can be interpreted as another indication for success of this surgery (Alexander, 1990). Ocular inflammation did not seem to have worsened the ERU-score further after the vitrectomy. The lens score increased significantly from a mean score of 1.5 to 2.8. Cataracts seemed to have an ongoing maturation and another study was able to show that 13 vitrectomized eyes had a progressive postsurgical cataract formation, 11 of which already showed incipient cataracts or vitreal inflammatory infiltrates adhered to the posterior capsule of the lens prior to the vitrectomy (Frühauf, 1998). The progressive nature of cataract formation in horses can be explained by alterations within the complex anatomy of the lens. If inflammatory debris, for example, was attached to the posterior lens capsule and removed within the vitrectomy, shear forces effected the hyaline membrane of the lens and the usually inactive lens epithelium might have been triggered to start mitosis and the newly formed cells might undergo hyaline degeneration, lipid metamorphosis and calcification (Cogan, 1962). In another study, two out of 53 horses showed a cataract two months after the vitrectomy (Von Borstel et al, 2005). The same authors described in a later study that the higher the preexisting ocular changes, the higher the odds for postsurgical complications such as retinal detachment and progressive cataract formation were (Von Borstel et al, 2010). If lenses showed cataract formations prior to the surgery, the chance of postoperative progressive cataract formation was found to be high.

Vitreal opacities due to inflammatory debris were the most common findings in eyes of horses with ERU (Bartel, 2004; Kulbrock, 2013; Wiehen, 2012). The mean score of all eyes examined for the vitreous was 2.5 prior to the vitrectomy in the current study and decreased significantly to 1.7 after the vitrectomy. This was the principle alteration of the ERU-score caused directly by the vitrectomy procedure.

The recurrent episodes of intraocular inflammation usually lead to a destruction of intraocular tissue. ERU has been described to be the most common cause of blindness in horses (Angelos et al, 1988; Lavach, 1990). A recent study evaluated the prognosis and impact of ERU in the United States and reported a high frequency of blindness as well (Gerding & Gilger, 2016). Vison has been reported to be maintained in 72 to 85 % of the eyes with ERU that underwent a vitrectomy (Frühauf, 1998; Winterberg & Gerhards, 1997). Winterberg et al. (1997) differentiated between an increase and decrease of vision after vitrectomy. According to that study, an increase of vision occurred in 39.5 % (17/43) of the eyes, whereas vision decreased in 32.6 % (14/43) of the eyes and 27.9 % of the eyes were blind after the vitrectomy in their study. Von Borstel et al. (2005) reported that 20 % (10/47) of the vitrectomized eyes showed a complete loss of vision post-vitrectomy. In the present study, a total of 87 % (128/147) of the reexamined eyes had a positive dazzle reaction and a positive menace response. Only 3 % (5/147) of the reexamined eyes showed a negative dazzle reaction and 10 % (14/147) of the eyes showed a negative menace response. It is not possible to state at what time point the eyesight impaired after the vitrectomy. Further prospective studies need to reexamine vitrectomized eyes at certain time points after the vitrectomy.

5.5.3 Gender

Previous studies suggest that male horses suffer from ERU significantly more often than females (Dorrego-Keiter E, 2016; Kulbrock et al, 2013b; Szemes & Gerhards, 2000; Wollanke, 1995). Geldings have been reported to be affected significantly more often than mares and stallions (Szemes & Gerhards, 2000). The present study verified these findings, because more male than female, and more geldings than mares and stallions were vitrectomized compared to the clinic population during the same period of time (Tab. 3). Wollanke et al. (1995) previously demonstrated that the Leptospiral antibody titers were similar for male and female horses. Hence, these authors postulated a protective action of female hormones. Other uveitic diseases in human medicine have been described to be gender-predisposed. Morbus Behçet syndrome and Fuch's heterochromic cyclitis are two diseases seen in human patients that involve uveitic symptoms. These diseases were thought to be gender-related (O'Brien et al., 1994), but recent studies have cast doubt whether gender influences do exist in these diseases (Bonfioli et al, 2005; Davatchi et al, 2012). Since the mean ERU-score pre- and post-vitrectomy did not alter significantly between male and female horses, a different outcome after the vitrectomy between male and female horses seems altogether be nonexistent in the present study.

5.5.4 Breed, coat color and age

In the current study, Icelandic and Frisian horses, Trotters, Pura Raza Espanola and Knabstruppers were significantly more often vitrectomized than Warmbloods, as compared to the clinic population (Tab. 3). This may well be due to a higher prevalence of ERU in these breeds. In a previous retrospective study over one third of all horses with ERU were Icelandic horses (Wiehen, 2012).

Appaloosas are significantly more likely to develop uveitis than Standardbreds and Thoroughbreds (McLaughlin et al, 1992). Dwyer et al. (Dwyer et al, 1995b) described that Appaloosas have an 8.3 times higher chance of developing uveitis than other equine breeds. One study evaluated the genetic risk factors for Appaloosas, such as a marker in the TRPM1 gene on ECA1 region responsible for their spotted coat color and two different markers in the region for the equine major histocompatibility complex (MHC) (Fritz et al, 2014). These authors were able to identify three markers in these regions that were significantly associated with ERU. A leopard complex spotting gene that is associated with a higher risk of ERU and congenital stationary night blindness in the horse was found in Knabstruppers (Baumgart & Gerhards, 2014; Bellone et al, 2013; Fritz et al, 2014). In the present study, all Knabstruppers showed a postsurgical REC. This may be due to the different subtype of ERU in this breed and perhaps even due to a solely autoimmune or genetically triggered uveitis. Therefore, a long-term sustained cyclosporine implant may be the better choice of therapy in this breed (Baumgart, 2014; Baumgart & Gerhards, 2014; Gilger & Allen, 1998; Gilger et al, 2000; Gilger et al, 2010). A genetic risk factor for developing ERU was also shown for Warmblood horses (Kulbrock et al, 2013a).

Since “insidious ERU” has been described as mild and painless, owners often do not recognize ocular changes until loss of vision occurs. “Insidious ERU” has less often been associated with an intraocular leptospiral infection in Appaloosa and Knabstruppers and a larger primarily immunological etiological component than in other breeds has, therefore, been assumed (Baumgart, 2014; Gilger et al, 2010).

The group with the coat color “other” was affected by REC significantly more frequently in the present study than horses with a “black & brown” coat color. One reason for this may be that all horses with a spotted coat color were assigned to this group and, as outlined above, “spotted” horses appear to have a poorer prognosis concerning their rate of REC than single-color horses. Horses with a black, brown and black-brown coat color (“black & brown”) were vitrectomized significantly more often in the current study than horses with the coat color “other” (Tab. 3). This is in accordance with the higher occurrence of ERU among Warmbloods with a “black & brown” coat color (Kulbrock et al, 2013b). Other studies have demonstrated an overrepresentation of horses with ERU of a brown coat color as well (Bartel, 2004; Wiehen, 2012).

A total of 45.1 % (106/235) of all horses vitrectomized were between four and seven years of age. This age group was significantly overrepresented compared to any other age group. Information about the prevalence of ERU among horses of different ages varies in literature. Some authors stated that relatively young horses were affected (Alexander, 1990; Wollanke, 1995) while others reported an increase of ERU among older horses, supposedly because of the cumulative contact to leptospiral antigens and antibodies during their lives (Bartel, 2004; Szemes & Gerhards, 2000). The current study implies that horses of an age between four and seven years are more likely to be vitrectomized. The reason for this may be that younger horses were more prone to develop ERU, that owners were more prone to invest in a vitrectomy in younger horses, or that referring veterinarians were able to diagnose ERU earlier than in older horses. Referring veterinarians may have recommended vitrectomy at an early stage of the disease, due to a better outcome at this stage (Von Borstel et al, 2010).

5.5.5 Leptospiral-status

One of the propagated etiological (infectious) factors for ERU in North-West Europe is an intraocular leptospiral infection. In previous studies vitreous and aqueous humor of affected eyes has been analyzed by different methods, such as PCR, MAT and ELISA, to identify a potential intraocular infection with leptospires (Alexander, 1990; B. Wollanke, 2004; Brandes et al, 2007; Brem et al, 1998; 1999; Faber et al, 2000; Frellstedt, 2009; Gerding & Gilger, 2016; Gilger et al, 2008; Halliwell, 1985; Heusser, 1948; Niedermaier et al, 2006; Pearce et al, 2007; Polle et al, 2014; Rimpau, 1947; Schwink et al, 1989; Sillerud et al, 1987; Von Borstel, 2010; Wiehen, 2012; Wollanke et al, 1998; Wollanke et al, 2000; Wollanke et al, 2001). In the present study, intraocular leptospiral infection was demonstrable through MAT and/or PCR in a total of 56 % of the vitrectomized eyes. This finding is in accordance with a recent study performed in northern Germany in which 221 undiluted vitreal samples were examined by MAT and culturing. The MAT revealed that 57.7 % of the samples were positive and 16 % of the samples had a positive leptospiral culture (Dorrego-Keiter E, 2016). It is still under discussion whether the prognosis is better in leptospiral-positive eyes after a vitrectomy. Tömördy et al. (2010) reported a higher number of eyes without recurrent episodes of uveitis if leptospiral antibodies were detectable. In contrast to this study, Dorrego-Keiter et al. (2017) demonstrated that there is no significant difference between the outcome of eyes tested negative and tested positive for leptospiral antibodies via MAT (Dorrego Keiter et al, 2017; Tömördy et al, 2010). All odds ratios calculated in the present study for eyes to have a REC depending upon being leptospiral-positive or leptospiral-negative showed a 95 % confidence interval below one. Hence, none of these odds ratios were significant, although the odds ratio for leptospiral-negative eyes to have a REC was found to be higher (OR: 1.925). When Warmbloods are compared to Icelandic horses, independent to their intraocular leptospiral status, Icelandic horses are 1.4 times more likely to develop REC. Therefore, not only the intraocular leptospiral status, but also the breed appeared to have an influence on the prognosis after vitrectomy. Since these odds ratios were only tendencies and 82 of the leptospiral-negative eyes did not show any signs of REC, vitrectomy did not appear to be significantly less effective on leptospiral-negative than on leptospiral-positive eyes. If only leptospiral-positive eyes would have been included, the rate of non-recurrence would be slightly higher for Warmblood horses (64/70, 91 %), but lower for Icelandic horses (15/19, 79 %). The findings of the present study, therefore, support the concept that the etiology of ERU is multifactorial, and other factors influencing the long-term prognosis after a vitrectomy most probably are auto immunogenic and genetically determined (Deeg, 2008; Deeg et al, 2001; Gilger, 2010; Kulbrock et al, 2013a; Von Borstel et al, 2010).

5.6 Conclusion

This study showed that between 85 % (74 % FOC, 11 % NOIS) of the evaluated eyes and 91 % (73 % FOC, 18 % NOIS) of the reexamined eyes showed an improvement (IMP) of the recurrent uveitic episodes after the vitrectomy. And a total of 87 % (128/147) of the reexamined eyes did show a positive dazzle reflex and a positive menace response after the vitrectomy at the time of ocular reexamination for the current study.

The odds of having a recurrence were higher if no leptospiral DNA or antibodies were detected. However, genetic and potentially autoimmunogenic factors, as well as age, breed, and preexisting ocular damage appears to influence the long-term prognosis. Therefore, an early vitrectomy of horses with a reliable diagnosis of classic ERU should be advocated.

6 Übergreifende Diskussion

Die ERU wird als eine multifaktorielle Erkrankung angesehen, bei der je nach Erkrankungsform unterschiedliche Pathogenesen im Mittelpunkt stehen. So wird die „intermediäre/schleichende Uveitis" der Appaloosa-Pferde mit genetischen Faktoren in Verbindung gebracht, aber auch ein autoimmuner Einfluss konnte dargelegt werden (Deeg, 2008; 2009; Deeg et al, 2002b; Fritz et al, 2014; Regan et al, 2012). Doch bis heute ist die intraokulare Leptospireninfektion der meist untersuchte pathogenetische Faktor der ERU. Intraokulare, leptospirale Antikörper konnten mit Hilfe des MATs in bis zu 80 % der betroffenen Pferdeaugen nachgewiesen werden (Faber et al, 2000; Frellstedt, 2009; Gerding & Gilger, 2016; Gilger, 2010; Gilger et al, 2008; Wiehen, 2012; Wollanke et al, 1998; Wollanke et al, 2004; Wollanke et al, 2000; Wollanke et al, 2001).

6.1 MAT

In Kapitel I („Intraokularer Leptospirennachweis mittels PCR, MAT, ELISA und Kultur von Pferden mit equiner rezidivierender Uveitis (ERU) in verschiedenen Laboren") wurde gezeigt, dass die Ergebnisse des Antikörpernachweises mittels MAT sich von Labor zu Labor rasse- und serovarabhängig unterscheiden. Leptospirale Antikörper werden serologisch mit Hilfe lokal-spezifischer leptospiraler Serovare anhand unterschiedlicher Agglutinationseigenschaften klassifiziert und im Rahmen des MAT wird die quantitative Agglutination möglicher Antikörper mit leptospiralen Strängen ausgewählter Serovare beurteilt (Kmety & Dikken, 1993). Dieses Vorgehen kann jedoch aufgrund von Kreuzagglutinationen zu Ungenauigkeiten führen (Kmety & Dikken, 1993; Turner, 1968). Die serovar- und rasseäbhängigen Unterschiede des MAT zwischen den Laboren in der hier vorliegenden Studie können möglicherweise auf den Gebrauch eng verwandter, jedoch unterschiedlicher leptospiraler Stränge des gleichen Serovars zurückgeführt werden. Die Ergebnisse des Kapitel I werden untermauert durch eine Studie, die identische Proben von Rindern und Schafen an verschiedene Labore zum Leptospiren-Nachweis mittels PCR und MAT schickte (Fang et al, 2014). Es konnte ebenfalls eine sehr hohe Übereinstimmung der PCR-Untersuchungen und eine serovarabhängige Übereinstimmung der MAT-Untersuchungen festgestellt werden. In einer weiteren Studie wurden weltweite Ringversuche durchgeführt, die gezeigt haben, dass Labore, die schon im ersten Ringversuch teilgenommen hatten insgesamt höhere Titer identischer Serovare mittels MAT diagnostizieren konnten als Labore, die nur am zweiten Ringversuch teilnahmen (Chappel et al, 2004). Somit kann sowohl die labordiagnostische Methode, als auch der Erfahrungswert der Labore selbst einen Einfluss auf die Ergebnisse eines Leptospirennachweises haben. Der Gebrauch unterschiedlicher Labore kann dementsprechend eine Erklärung für die divergierenden MAT-Ergebnisse in der Literatur sein.

6.2 PCR

Die Ergebnisse der PCR-Untersuchungen zeigten, trotz unterschiedlicher Versuchsmethodik und unterschiedlicher leptospiraler Zielgene, sehr gute Übereinstimmungen, die zudem auch rasse- und serovarunabhängig waren. Eine Erklärung für diese sehr gute Übereinstimmung ist, dass trotz unterschiedlicher PCR-Methodik und unterschiedlicher Zielgene beide PCR-Protokolle eine ähnlich hohe Sensitivität hatten. Die hohe Sensitivität von verschiedenen PCR-Protokollen wurde in unterschiedlichen Studien mit Hilfe von Verdünnungsreihen gezeigt. In eine dieser Studien wurde das leptospirale rrs(16S)-Gen amplifiziert und ein positives PCR-Ergebnis konnte ab einem Gehalt von zwei bis zehn Bakterien pro Milliliter Probenmaterial beobachtet werden (Merien et al, 1992). In einer anderen Studie wurde mit Hilfe einer Real-

Time PCR das leptospirale lfb1-, secY- und lipL32-Gen amplifiziert und ein positives Ergebnis dieser PCR wurde ab einem Gehalt von einem Bakterium beziehungsweise fünf Genomäquivalenten pro Mikroliter Probenmaterial erreicht (Bourhy et al, 2011). Dementsprechend ist davon auszugehen, dass nur eine sehr starke Probenverdünnung beziehungsweise ein sehr geringer Gehalt an leptospiralen Antigen die Sensitivität der PCR beeinflussen.

6.3 Vitrektomie

Der Austausch des uveitisch veränderten Glaskörpers im Rahmen einer Vitrektomie ist eine mögliche Therapieoption der ERU. Es ist jedoch unklar, welche Faktoren den Erfolg dieser Therapie beeinflussen. In einer Studie mit 47 an ERU erkrankten Augen wurde der Einfluss eines positiven Leptospirennachweises aus Glaskörper mittels MAT retrospektiv überprüft. Augen, in denen der Antikörpernachweis negativ verlief zeigten häufiger Rezidive nach der Vitrektomie (6/7), als Augen, aus deren Glaskörper leptospirale Antikörper nachgewiesen werden konnten (40/47) (Tömördy et al, 2010). In Kapitel II („Langzeit-Verlaufskontrolle von an equiner rezidivierender Uveitis (ERU) erkrankten Pferden nach der Vitrektomie") konnte kein signifikanter Einfluss des positiven Leptospirennachweises erbracht werden. Augen, aus deren Glaskörper leptospirale Antikörper oder / und leptospirales Antigen nachgewiesen werden konnte, zeigten in der vorliegenden Studie nicht signifikant weniger Rezidive (13/121), als Augen bei denen diese nicht nachgewiesen werden konnten (19/101). Das relative Risiko ein Rezidiv zu erleiden ist bei Augen aus deren Glaskörpermaterial weder leptospirales Antigen, noch leptospirale Antikörper nachgewiesen werden konnten zwar um den Faktor 1,9 höher als bei Augen mit einem positivem Leptospirennachweis, jedoch ist dieses relative Risiko nicht signifikant höher. Außerdem bleiben der Einfluss der ausgewählten labordiagnostischen

Nachweismethode, sowie der Einfluss der ausgewählten Labore, wie in Kapitel I erläutert, bestehen. Zusammenfassend ergibt sich somit aus den beiden Studien, dass für die Entscheidung zur Vitrektomie als Behandlungsmethode nicht zwingend ein positiver Leptospirennachweis zu fordern ist.

6.4 Leptospiren, Genetik

Grundsätzlich können sowohl im Serum von an ERU erkrankten, als auch bei augengesunden Pferden in gleichem Maße Antikörper gegen Leptospiren nachgewiesen werden (Wollanke et al, 1998; Wollanke et al, 2004). Es ist bisher jedoch nicht gelungen zu klären, wie und aus welchen Gründen die Leptospiren selbst oder leptospirale Antikörper bei manchen Pferden in das Auge eindringen können. Es gibt Vermutungen, dass genetische Variationen bei an ERU erkrankten Pferden eine Rolle spielen. So konnte laut zum Beispiel ein Rückgang der ERU-Erkrankungen festgestellt werden, nachdem erkrankte Pferde von der Nachzucht ausgeschlossen wurden (Alexander, 1990; Cross, 1966). Des Weiteren konnte in einer Studie mit 69 Warmblutpferden bei 13 von 32 an ERU erkrankten Pferden das ELA (equine lymphocyte antigen) Haplotyp A9 nachgewiesen werden, während bei keinem der augengesunden Kontrollpferde diese Ausprägung gefunden werden konnte (Deeg et al, 2004). In einer anderen Studie mit deutschen Warmblutpferden konnte gezeigt werden, dass ein signifikanter Einzelnukleotid-Polymorphismus auf dem equinen Chromosom 20 in direkter Nachbarschaft zu IL-17A und IL-17F, bei erkrankten Pferden vorlag (Kulbrock et al, 2013a). Diese beiden Gene könnten somit ein entscheidender Unterschied zwischen augengesunden und erkrankten Pferden sein. Ob es alleinig aufgrund dieser genetischen Unterschiede zu einer ERU kommen kann, oder ob es zu einer veränderten Reaktion auf eine Leptospireninfektion kommt, bleibt jedoch unklar.

Bei Pferden der Rasse Appaloosa konnten drei Marker identifiziert werden, die signifikant mit der „schleichenden Uveitis“ in Zusammenhang gebracht wurden. Ein Einzelnukleotid-Polymorphismus korrelierte sehr stark mit dem TRPM1-Gen der ECA1-Region, der Genort des Tigerschecken-Komplexes. Die anderen beiden Marker standen in Verbindung mit dem Haupthistokompatibilitätkomplex (MHC) in der ECA20-Region (Fritz et al, 2014). Ebenso konnte für die Rasse Knabstrupper ein Tigerschecken-Komplex-Gen ermittelt werden, das mit einem erhöhten Risiko für kongenitale stationäre Nachtblindheit und ERU in Verbindung gebracht wird (Alessandra Baumgart & Gerhards, 2014; Bellone et al, 2013; Fritz et al, 2014). Im Kapitel II der vorliegenden Arbeit konnte gezeigt werden, dass die Fellfarbe und die Rasse einen signifikanten Einfluss auf die Rezidivfreiheit nach der Vitrektomie hat. So zeigten Pferde mit einer anderen Fellfarbe als fuchs, braun, dunkelbraun oder schwarz ein 2,6-Fach höheres Risiko ein Rezidiv zu erleiden, was signifikant ist. Die Rasse der Knabstrupper zeigte in der vorliegenden Arbeit ebenfalls signifikant häufiger Rezidive, als alle anderen Rassen. Das Risiko eines Rezidivs ist für Islandpferde im Vergleich zu Warmblutpferden um ein 1,5-Faches höher. Schlussfolgernd ist eine Vitrektomie bei bestimmter Farbausprägung beziehungsweise bei bestimmten Rassen weniger erfolgreich, als bei anderen. Dies verdeutlicht eine nicht zu vernachlässigende genetische Komponente der ERU oder sogar auf einen genetischen Ursprung.

6.5 Autoimmune Komponenten

Neben genetischen Komponenten und einer intraokularen Leptospireninfektion wird die ERU auch immer wieder als eine autoimmune Erkrankung eingeordnet. Gestützt wird diese Theorie grundsätzlich durch den wiederkehrenden Charakter der ERU, den erfolgreichen Einsatz immunmodulatorischer Medikamente und den unbeständigen Erfolg antibiotischer Therapien

(Deeg et al, 2002a; Regan et al, 2012). So konnte gezeigt werden, dass an ERU erkrankte Augen hauptsächlich von $CD4^+$-T-Lymphozyten infiltriert sind, die eine Th1-Immunantwort durch Interleukin-2 (IL-2) und Interferon-ɣ (IFNɣ) auslösen (Deeg et al, 2002a; Deeg et al, 2001; Gilger et al, 1999; Romeike et al, 1998). In weiteren Studien wurden Autoantigene identifiziert, die Autoreaktionen von B- und T-Lymphozyten auslösen. Das Interphotorezeptor-Retinoid-Bindungs-Protein (IRBP) und das Zelluläre-Retinaldehyd-Bindungs-Protein (CRALBP) waren in diesen Studien in der Lage nach subkutanen Injektionen eine stark der ERU ähnelnde Uveitis bei Pferden auszulösen (Deeg, 2009; Deeg et al, 2007; Deeg et al, 2006; Deeg et al, 2002b). Im Zusammenhang mit einer intraokularen Leptospireninfektion konnten ebenfalls Rückschlüsse auf immunologische Komponenten der ERU gezogen werden (Wollanke et al, 2004). So konnte gezeigt werden, dass die leptospiralen Lipoproteine LruA und LruB immunologisch relevante Epitope mit der equinen Linse und Retina teilen (Verma et al, 2005; Verma et al, 2010). Bis heute ist unklar, ob eine intraokulare Leptospireninfektion die alleinige Ursache dieser immunologischen Komponenten der ERU darstellt, oder ob es eine bisher noch nicht klinisch eindeutig differenzierte Form der ERU gibt, die eine echte Autoimmunreaktion darstellt (Wollanke et al, 2004).

6.5 Alter

Einige Autoren fanden heraus, dass das Alter mit der Häufigkeit der ERU korreliert und sie gehen davon aus, dass ältere Pferde im Laufe ihres Lebens häufiger Kontakt mit Leptospirenantigen haben und die ERU als Folge einer Leptospireninfektion auftritt (Bartel, 2004; Szemes & Gerhards, 2000). Im Gegensatz dazu wurde im Kapitel II der vorliegenden Arbeit deutlich, dass der Leptospirennachweis aus Glaskörpermaterial bei älteren Pferden weniger häufig positiv ausfällt. Unterschiedliche Ansätze können diese Beobachtung erklären.

Entweder waren leptospirales Antigen und/ oder Antikörper zu keinem Zeitpunkt vorhanden, sie waren zum Zeitpunkt der Vitrektomie nicht mehr vorhanden, oder die verwendeten Nachweismethoden waren nicht in der Lage leptospirales Antigen oder Antikörper nachzuweisen. Da im Kapitel I gezeigt werden konnte, dass die Kombination aus einer PCR- und einer MAT-Untersuchung nahezu alle Leptospiren-positiven Glaskörperproben detektierte, die von einer Kombination aus kultureller Anzucht, doppelter PCR- und doppelter MAT-Untersuchung nachgewiesen werden konnten, ist von einem Versagen der Nachweismethoden bei älteren Pferden im Kapitel II nicht auszugehen. Dass leptospirales Antigen und/ oder Antikörper zum Zeitpunkt der Vitrektomie nicht mehr vorhanden ist, spricht gegen die Theorie der persistierenden Leptospireninfektion des Auges (Wollanke et al, 2001). Dass zum Zeitpunkt der Vitrektomie weder leptospirales Antigen, noch leptospirale Antikörper nachweisbar waren, kann jedoch ein Hinweis auf eine erfolgreiche Elimination der Leptospiren aus dem Auge sein. Somit könnten die rezidivierenden Schübe wiederum auf eine durch eine intraokulare Leptospireninfektion induzierte, immunologisch vermittelte Antwort hindeuten (Gilger et al, 2008). Bisher gibt es keine Studien die dies eindeutig belegen können.

Im Hinblick auf die vorliegenden Ergebnisse des Kapitel I und II lässt sich keine Empfehlung bezüglich einer präoperativen Kammerwasseruntersuchung auf leptospirale Antikörper mittels MAT aussprechen. Vielmehr sollten Faktoren wie Rasse, Farbe, Alter und Vorschädigungen des betroffenen Auges für eine Entscheidung pro oder kontra Vitrektomie in Betracht gezogen werden.

7 Zusammenfassung

Eva Baake:

Studie zur Ursache und Therapie der equinen rezidivierenden Uveitis

Die equine rezidivieredne Uveitis (ERU) ist eine Augenerkrankung, die vor allem durch wiederkehrende Entzündungsschübe verschiedener Augenabschnitte geprägt ist. Es gibt verschiedene äthiologische Theorien der ERU, von denen die „klassische ERU" bis heute am häufigsten mit einer intraokularen Leptospireninfektion in Verbindung gebracht wird. Die Pars plana Vitrektomie ist derzeit eine vielversprechende Therapieoption und es gibt Hinweise darauf, dass diese Behandlungsmethode erfolgreicher ist, wenn intraokulare Antikörper gegen Leptospiren nachweisbar sind. Aus diesem Grund wird eine Kammerwasserpunktion betroffener Augen vor der Vitrektomie empfohlen. Der Nachweis einer intraokularen Infektion mit Leptospiren wird durch die Untersuchung von Kammerwasser oder Glaskörpermaterial aus betroffenen Augen erbracht. Gängige Nachweismethoden sind der Mikroagglutinationstest (MAT) zum Nachweis intraokularer Antikörper gegen Leptospiren, sowie die Polymerase-Kettenreaktion (PCR) zum Nachweis leptsospiraler DNA und die kulturelle Anzucht zum Nachweis vermehrungsfähiger Leptospiren. Die Nachweisrate aus intraokularem Material betroffener Augen mittels MAT, PCR und Kultur variiert in der Literatur deutlich.

In Teil 1 der vorliegenden Arbeit „Detection of intraocular leptospiral DNA, antibodies and Leptospira spp. in horses with equine recurrent uveitis in different laboratoriers" wird der Einfluss unterschiedlicher Labore auf Untersuchungsergebnisse und Nachweisraten identischer Glaskörperproben untersucht. Es wurden Glaskörperproben von 78 Pferden (93 Augen) zu Beginn der Vitrektomie gewonnen. Diese wurden gleichmäßig geteilt und an drei verschiedene

Labore (L1, L2 und L3) in Deutschland versandt. Labor 1 und 2 untersuchten die Proben auf leptospirales Antigen mittels PCR, L1 und L3 untersuchten sie auf leptospirale Antikörper mittels MAT und L3 führte eine kulturelle Anzucht der Proben durch. Die Nachweisrate leptospiraler DNA lag bei 38 % (53/85) mittels PCR in L1 und L2. Die Übereinstimmung beider Labore lag bei 98%, lediglich zwei Proben erbrachten gegensätzliche Ergebnisse. Ein positiver Antikörpernachweis mittels MAT war im L1 bei 51% (45/88) der Proben und im L3 bei 41% (36/88) der Proben erfolgreich. Insgesamt zeigte der MAT eine Übereinstimmung der Ergebnisse von 71 %. Zusätzlich war die Übereinstimmung der MAT-Ergebnisse aus L1 und L3 serovar- und rasseabhängig. Die kulturelle Anzucht war in 16 % (12/75) der Proben erfolgreich. Insgesamt durchliefen 67 Glaskörperproben parallel alle Untersuchungen (PCR, MAT, Kultur) in allen drei Laboren (L1, L2, L3). Der Anteil an Glaskörperproben, die in mindestens einer Untersuchung ein positives Ergebnis hatten lag bei 57 % (38/67).

In Teil 2 der vorliegenden Arbeit „Long-term ophthalmologic examinations of eyes with equine recurrent uveitis after pars plana vitrectomy“ wird eine Langzeit-Verlaufskontrolle von an ERU erkrankten Pferden nach der Vitrektomie vorgestellt und die Ergebnisse in Kontext mit Faktoren wie Alter, Geschlecht, Rasse, Fellfarbe und Vorschädigungen der vitrektomierten Augen gesetzt. Es konnten 246 von 284 vitrektomierte Augen in einem Zeitraum von 17 bis 123 Monaten (Mittelwert: 64 Monate) nach der Vitrektomie evaluiert werden. Eine ophthalmologische Nachuntersuchung konnte an 147 Augen durchgeführt werden. Insgesamt zeigten 182 (74 %) der 246 evaluierten Augen im Zeitraum zwischen der Vitrektomie und der Evaluierung keine erneuten Entzündungsschübe des operierten Auges, wohingegen 27 (11 %) Augen unspezifische entzündliche Symptome zeigten und 37 (15 %) Augen nach der Vitrektomie entzündliche Schübe der gleichen Art, wie vor der Vitrektomie zeigten

(„Rezidiv“). Die Anzahl der Schübe vor der Vitrektomie, das Geschlecht der vitrektomierten Pferde und die Ergebnisse des Leptospirennachweises mittels MAT und PCR aus Glaskörpermaterial hatten keinen signifikanten Einfluss auf das Auftreten von Rezidiven. Pferde mit einer anderen Fellfarbe als fuchs, braun, rappe und dunkelbraun zeigten signifikant häufiger Rezidive, und Pferde mit einem ERU-Score von 5 und 3 zeigten signifikant häufiger Rezidive als Pferde mit einem ERU-Score von 2. Außerdem zeigten Pferde der Rasse „Knabstrupper“ signifikant häufiger Rezidive als alle anderen Rassen. Weiterhin konnte gezeigt werden, dass die meisten Rezidive in einem Zeitraum von zwei Monaten nach der Vitrektomie stattfanden (17/37). Dennoch trat auch nach bis zu 123 Monaten nach der Vitrektomie ein Rezidiv auf. Die Rezidivfreiheit nach der Vitrektomie lag für Warmblutpferde bei 88 %.

Im Hinblick auf die vorliegenden Ergebnisse des Kapitel I und II lässt sich keine Empfehlung bezüglich einer präoperativen Kammerwasseruntersuchung auf leptospirale Antikörper mittels MAT aussprechen. Vielmehr sollten Faktoren wie Rasse, Farbe, Alter und Vorschädigungen des betroffenen Auges für eine Entscheidung pro oder kontra Vitrektomie in Betracht gezogen werden.

8 Summary

Eva Baake

Study on cause and therapy of equine recurrent uveitis (ERU)

Equine recurrent uveitis (ERU) is an ocular disease characterized mainly by recurrent inflammatory episodes of different segments of the eye. There are numerous aetiological theories for the cause of ERU, but to date, the "classic ERU" is most commonly associated with an intraocular leptospiral infection. The pars plana vitrectomy has been the most promising treatment of ERU and is thought to be more successful in eyes with detectable leptospiral antibodies in the microscopic agglutination test (MAT) and therefore a preselection is recommended.

Vitreous material or aqueous fluid is examined by the MAT to detect leptospiral antibodies, the polymerase chain reaction (PCR) is used to detect leptospiral antigen and a cultural examination can detect living leptospires. The detection rate of ocular material from affected eyes is varying widely in Literature. The article "Detection of intraocular leptospiral DNA, antibodies and Leptospira spp. in horses with equne recurrent uveitis in different laboratories" examines the impact of different laboratories on the MAT- and PCR-results of the same vitreal samples. Vitreal samples were obtained from 93 eyes of 78 horses at the beginning of the vitrectomy. These samples were evenly split and sent to three different laboratories (L1, L2 and L3) in Germany. L1 and L2 performed a PCR, L1 and L3 carried out a MAT and L3 performed a cultural examination. The detection-rate of leptospiral DNA by PCR was 38 % in L1 and L2. The accordance was therefore 98 %, only two specimens showed deviating results. The MAT in L1 was positive in 51 % (45/88) of the vitreal samples and it was positive in L3 in 41 %

(36/88) of the samples. The conformity was 71 % for L1 and L3, and it was moreover serovar- and breed-dependent. The cultivation of Leptospires was successful in 16 % (12/75) of the vitreal specimens. Overall, 67 vitreal samples were tested in all three laboratories. A positive result in at least one of the tests performed occurred in 57 % (38/67) of the samples.

The article "Long-term ophthalmologic examinations of eyes with equine recurrent uveitis after pars plana vitrectomy" gives a detailed prognosis for horses after the vitrectomy, based on long-term evaluations in context to age, gender, breed, coat colour and ocular damage prvious to the vitrectomy. A total of 246 of 284 eyes that underwent pars plana vitrectomy could be evaluated between 17 to 123 months (⍰: 64 months) after the surgery. An ophthalmologic examination was performed in 147 eyes. Altogether, 182 (74 %) of the 246 evaluated eyes did not show any signs of ocular inflammation after the vitrectomy. Whereas 27 (11%) eyes had non-specific ocular inflammatory symptoms and 37 (15 %) eyes showed recurrences of the same magnitude as prior to the vitrectomy. The number of episodes prior to the vitrectomy, the gender of the horse, and the leptospiral status of the eye showed no significant influence on the amount of recurrences. Horses with a coat colour other than chestnut, black, brown and black-brown had significantly more recurrences and horses with an ERU-score of 5 and 3 had significantly more recurrences than horses with an ERU-Score of 2. Furthermore horses of the breed "Knabstrupper" showed significantly more recurrences than every other breed. Most of the recurrences occurred in a period of two months after the vitrectomy (17/37). Nevertheless, recurrences were seen up to 123 months after the vitrectomy. The rate for warmblood horses without recurrences was 88 %.

9 Literaturverzeichnis

Ahmed, N., Devi, S. M., Valverde Mde, L., Vijayachari, P., Machang'u, R. S., Ellis, W. A. & Hartskeerl, R. A. (2006):
Multilocus sequence typing method for identification and genotypic classification of pathogenic Leptospira species.
Annals of Clinical Microbiology and Antimicrobials, Nov 23; 5: 28.

Alexander, C.-S. & Keller, H. (1990):
Ätiologie und Vorkommen der periodischen Augenentzündung des Pferdes im Raum Berlin.
Tierärztliche Praxis, 18, 623-627.

Angelos, J., Oppenheim, Y., Rebhun, W., Mohammed, H. & Antczak, D. F. (1988):
Evaluation of breed as a risk factor for sarcoid and uveitis in horses.
Animal Genetics, 19 (4), 417-25.

Wollanke, B., Brem, S., Meyer, P., Forbrig, T., Grassl, P., Gerhards, H.; & Kopp, H. (2004):
Prophylaxe der equinen rezidivierenden Uveitis (ERU): Erste Erfahrungen mit einem Leptospiren-Impfstoff bei Pferden.
Pferdeheilkunde, 20 (5), 447-454.

Bartel, S. F. (2004):
Retrospektive kasuistische Analyse von 369 Pferden mit equiner rezidivierender Uveitis (ERU)
(Doctorial dissertation, LMU München).

Baumgart, A. (2014):
Cyclosporin A und dessen möglicher Einsatz bei der Tigerschecken-Uveitis.
(Doctorial dissertation, LMU München).

Baumgart, A. & Gerhards, H. (2014):
Besonderheiten der Tigerschecken-Uveitis und möglicher Cyclosporin A-Einsatz in der Therapie in Deutschland.
Pferdeheilkunde, 30 (6), 626 - 632.

Bellone, R. R., Holl, H., Setaluri, V., Devi, S., Maddodi, N., Archer, S., Sandmeyer, L., Ludwig, A., Foerster, D., Pruvost, M., Reissmann, M., Bortfeldt, R., Adelson, D. L. & Brooks, S. A. (2013):
Evidence for a retroviral insertion in TRPM1 as the cause of congenital stationary night blindness and leopard complex spotting in the horse.
PloS One. 8 (10), e78280.

Bonfioli, A. A., Curi, A. L. & Orefice, F. (2005):
Fuchs' heterochromic cyclitis.
Seminars in ophthalmology, 20 (3), 143-146. Taylor & Francis.

von Borstel, M., von Oppen, T., Glitz, F., Frühauf, B., Deegen, E., Boeve, M. H. & Ohnesorge, B. (2005):
Langzeitergebnisse der Pars plana Vitrektomie (double port) bei equiner rezidivierender Uveitis.
Pferdeheilkunde, 21 (1), 13-18.

von Borstel, M., Oey, L., Strutzberg-Minder, K., Boevé, M. H. & Ohnesorge, B. (2010):
Direkter und indirekter Nachweis von Leptospiren aus Glaskörperproben von Pferden mit ERU.
Pferdeheilkunde, 26(2), 219-225.

Bourhy, P., Bremont, S., Zinini, F., Giry, C. & Picardeau, M. (2011):
Comparison of real-time PCR assays for detection of pathogenic Leptospira spp. in blood and identification of variations in target sequences.
Journal of Clinical Microbiology, 49 (6), 2154-60.

Brandes, K., Wollanke, B., Niedermaier, G., Brem, S. & Gerhards, H. (2007):
Recurrent uveitis in horses: vitreal examinations with ultrastructural detection of leptospires.
Transboundary and Emerging Diseases, 54 (5), 270-275.

Brem, S., Gerhards, H., Wollanke, B., Meyer, P. & Kopp, H. (1998):
Demonstration of intraocular leptospira in 4 horses suffering from equine recurrent uveitis (ERU).
Berliner und Münchener Tierärztliche Wochenschrift, 111 (11-12), 415-7.

Brem, S., Gerhards, H., Wollanke, B., Meyer, P. & Kopp, H. (1999):
35 leptospira isolated from the vitreous body of 32 horses with recurrent uveitis (ERU).
Berliner und Münchener Tierärztliche Wochenschrift, 112 (10-11), 390-3.

Chappel, R. J., Goris, M., Palmer, M. F. & Hartskeerl, R. A. (2004):
Impact of proficiency testing on results of the microscopic agglutination test for diagnosis of leptospirosis.
Journal of Clinical Microbiology, 42 (12), 5484-8.

Cogan, D. G. (1962):
Anatomy of lens and pathology of cataracts.
Experimental Eye Research, 1 (4), 291-5.

Cook, C. S., Peiffer, R. L. & Harling, D. E. (1983):
Equine recurrent uveitis.
Equine Veterinary Journal, 15 (S2), 57-60.

Cross, R. S. (1966):
Equine periodic ophthalmia.
Veterinary Record, 78 (1), 8-13.

Curling (2011):
Equine recurrent uveitis: Treatment.
Compendium on Continuing Education for Veterinarians, 33 (6), E1.

Davatchi, F., Shahram, F., Chams-Davatchi, C., Sadeghi Abdollahi, B., Shams, H., Nadji, A., Faezi, T., Akhlaghi, M., Ghodsi, Z., Larimi, R. & Ashofteh, F. (2012):
Behcet's disease: is there a gender influence on clinical manifestations?
International Journal of Rheumatic Diseases, 15 (3), 306-14.

Deeg, C. A. (2008):
Ocular immunology in equine recurrent uveitis.
Veterinary Ophthalmology, 11 (Sl), 61-5.

Deeg, C. A. (2009):
A proteomic approach for studying the pathogenesis of spontaneous equine recurrent uveitis (ERU).
Veterinary Immunology and Immunopathology, 128 (1), 132-6.

Deeg, C. A., Altmann, F., Hauck, S. M., Schoeffmann, S., Amann, B., Stangassinger, M. & Ueffing, M. (2007):
Down-regulation of pigment epithelium-derived factor in uveitic lesion associates with focal vascular endothelial growth factor expression and breakdown of the blood-retinal barrier.
Proteomics, 7 (9), 1540-8.

Deeg, C. A., Ehrenhofer, M., Thurau, S. R., Reese, S., Wildner, G. & Kaspers, B. (2002a):
Immunopathology of recurrent uveitis in spontaneously diseased horses.
Experimental Eye Research, 75 (2), 127-33.

Deeg, C. A., Kaspers, B., Gerhards, H., Thurau, S. R., Wollanke, B. & Wildner, G. (2001):
Immune responses to retinal autoantigens and peptides in equine recurrent uveitis.
Investigative Ophthalmology & Visual Science, 42 (2), 393-8.

Deeg, C. A., Marti, E., Gaillard, C. & Kaspers, B. (2004):
Equine recurrent uveitis is strongly associated with the MHC class I haplotype ELA-A9.
Equine Veterinary Journal, 36 (1), 73-5.

Deeg, C. A., Pompetzki, D., Raith, A. J., Hauck, S. M., Amann, B., Suppmann, S., Goebel, T. W., Olazabal, U., Gerhards, H., Reese, S., Stangassinger, M., Kaspers, B. & Ueffing, M. (2006):
Identification and functional validation of novel autoantigens in equine uveitis.
Molecular & Cellular Proteomics, 5 (8), 1462-70.

Deeg, C. A., Thurau, S. R., Gerhards, H., Ehrenhofer, M., Wildner, G. & Kaspers, B. (2002b):
Uveitis in horses induced by interphotoreceptor retinoid-binding protein is similar to the spontaneous disease.
European Journal of Immunology, 32 (9), 2598-606.

Dorrego Keiter, E., Tóth, J., Dikker, L., Sielhorst, J. & Schusser, G. F. (2017):
Langzeitergebnisse der Pars-Plana-Vitrektomie in Abhängigkeit vom Leptospiren-Antikörper-Nachweis im Glaskörper bei 118 Pferden mit Equiner Rezidivierender Uveitis (ERU).
Pferdeheilkunde, 33 (2), 112-8.

Dorrego-Keiter E, Tóth, J., Dikker L, Sielhorst J, & Schusser G. F. (2016):
Detection of leptospira by culture of vitreous humor and detection of antibodies against leptospira in vitreous humor and serum of 225 horses with equine recurrent uveitis.
Berliner und Münchener Tierärztliche Wochenschrift, 129 (5-6), 209-15.

Dwyer, A. E., Crockett, R. S. & Kalsow, C. M. (1995a):
Association of leptospiral seroreactivity and breed with uveitis and blindness in horses: 372 cases (1986-1993).
Journal of the American Veterinary Medical Association, 207 (10), 1327-31.

Dwyer, A. E., Crockett, R. S. & Kalsow, C. M. (1995b):
Association of leptospiral seroreactivity and breed with uveitis and blindness in horses: 372 cases (1986-1993).
Journal of the American Veterinary Medical Association, 207 (10), 1327-31

Faber, N. A., Crawford, M., LeFebvre, R. B., Buyukmihci, N. C., Madigan, J. E. & Willits, N. H. (2000):
Detection of Leptospira spp. in the Aqueous Humor of Horses with Naturally Acquired Recurrent Uveitis.
Journal of Clinical Microbiology, 38 (7), 2731-3.

Fang, F., Collins-Emerson, J. M., Heuer, C., Hill, F. I., Tisdall, D. J., Wilson, P. R. & Benschop, J. (2014):
Interlaboratory and between-specimen comparisons of diagnostic tests for leptospirosis in sheep and cattle.
Journal of Veterinary Diagnostic Investigation, 26 (6), 734-47.

Frellstedt, L. (2009):
Equine recurrent uveitis: A clinical manifestation of leptospirosis.
Equine Veterinary Education, 21 (10), 546-52.

Fritz, K. L., Kaese, H. J., Valberg, S. J., Hendrickson, J. A., Rendahl, A. K., Bellone, R. R., Dynes, K. M., Wagner, M. L., Lucio, M. A., Cuomo, F. M., Brinkmeyer-Langford, C. L., Skow, L. C., Mickelson, J. R., Rutherford, M. S. & McCue, M. E. (2014):
Genetic risk factors for insidious equine recurrent uveitis in Appaloosa horses.
Animal Genetics, 45 (3), 392-9.

Frühauf, B., Ohnesorge, B., Deegen, E. & Boevé, M. (1998):
Surgical Management of equine recurrent uveitis with single port pars plana vitrectomy.
Veterinary Ophthalmology, 1 (2-3), 137-51.

Gerding, J. C. & Gilger, B. C. (2016):
Prognosis and impact of equine recurrent uveitis.
Equine Veterinary Journal, 48 (3), 290-8.

Gerhards, H. & Wollanke, B. (2001):
Uveitis bei Pferden - Diagnose und Therapie.
Pferdeheilkunde, 17 (4), 319-29.

Gerhards, H., Wollanke, B. & Brem, S. (1999):
Vitrectomy as a diagnostic and therapeutic approach for equine recurrent uveitis (ERU).
Proceedings of the Annual Convention of the AAEP, 45, 89-93.

Gilger, B. C. (2010):
Equine recurrent uveitis The viewpoint from the USA.
Equine Veterinary Journal, 42 (S37), 57-61.

Gilger, B. C. & Allen (1998):
Cyclosporine A in veterinary ophthalmology.
Veterinary Ophthalmology, 1 (4), 181-7.

Gilger, B. C., Malok, E., Cutter, K. V., Stewart, T., Horohov, D. W. & Allen, J. B. (1999):
Characterization of T-lymphocytes in the anterior uvea of eyes with chronic equine recurrent uveitis.
Veterinary Immunology and Immunopathology, 71 (1), 17-28.

Gilger, B. C., Malok, E., Stewart, T., Ashton, P., Smith, T., Jaffe, G. J. & Allen, J. B. (2000):
Long-term effect on the equine eye of an intravitreal device used for sustained release of cyclosporine A.
Veterinary Ophthalmology, 3 (2-3), 105-110.

Gilger, B. C. & Michau, T. M. (2004):
Equine recurrent uveitis: new methods of management.
Veterinary Clinics of North America: Equine Practice, 20 (2), 417-27.

Gilger, B. C., Salmon, J. H., Na, Y. Y., Barden, C. A., Chandler, H. L., Wendt, J. A. & Colitz, C. M. (2008):
Role of bacteria in the pathogenesis of recurrent uveitis in horses from the southeastern United States.
American Journal of Veterinary Research, 69 (10), 1329-35.

Gilger, B. C., Wilkie, D. A., Clode, A. B., McMullen, R. J., Utter, M. E., Komaromy, A. M., Brooks, D. E. & Salmon, J. H. (2010):
Long-term outcome after implantation of a suprachoroidal cyclosporine drug delivery device in horses with recurrent uveitis.
Veterinary Ophthalmology, 13 (5), 294-300.

Halliwell, R. E., Brim, T. A., Hines, M. T., Wolf, D. &White, F.H. (1985):
Studies on equine recurrent uveitis. II The role of infection with Leptospira interrogans serovar pomona.
Current Eye Research, 4 (10), 1033-40.

Hartskeerl, R. A., Goris, M. G. A., Brem, S., Meyer, P., Kopp, H., Gerhards, H. & Wollanke, B. (2004):
Classification of Leptospira from the Eyes of Horses Suffering from Recurrent Uveitis.
Zoonoses and Public Health, 51 (3), 110-15.

Heusser, H. (1948):
Die periodische Augenentzündung, eine Leptospirose?
Schweizer Archiv für Tierheilkunde, 90 (6), 287-314.

Kmety, E. & Dikken, H. (1993):
Classification of the species Leptospira interrogans and history of its serovars
University Press Groningen.

Kulbrock, M., Lehner, S., Metzger, J., Ohnesorge, B. & Distl, O. (2013a):
A genome-wide association study identifies risk loci to equine recurrent uveitis in German warmblood horses.
PLoS One, 8 (8), e71619.

Kulbrock, M., von Borstel, M., Rohn, K., Distl, O. & Ohnesorge, B. (2013b):
Studie zu Häufigkeit und Schweregrad der Equinen Rezidivierenden Uveitis bei Warmblütern.
Pferdeheilkunde, 29 (1), 27-36.

Landis, J. R. & Koch, G. G. (1977):
The measurement of observer agreement for categorical data.
Biometrics, 33 (1), 159-74.

Lavach, J. D. (1990):
Large animal ophthalmology. St. Louis: Mosby.

McLaughlin, S. A., Whitley, R. D. & Gilger, B. C. (1992):
Diagnosis and treatment of lens diseases.
Veterinary Clinics of North America: Equine Practice, 8 (3), 575-85.

McMullen Jr, R. (2015):
Intravitreal injection of low-dose gentamicin in horses for treatment of chronic recurrent or persistent uveitis: preliminary results.
Conference Paper, International Equine Ophthalmology Consortium/Acrivet, Inc. Symposium

Merien, F., Amouriaux, P., Perolat, P., Baranton, G. & Saint Girons, I. (1992):
Polymerase chain reaction for detection of Leptospira spp. in clinical samples.
Journal of Clinical Microbiology, 30 (9), 2219-24.

Niedermaier, G., Wollanke, B., Hoffmann, R., Brem, S. & Gerhards, H. (2006):
Detection of leptospira in the vitreous body of horses without ocular diseases and of horses with equine recurrent uveitis (ERU) using transmission-electron microscopy.
Deutsche Tierärztliche Wochenschrift, 113 (11), 418-22.

O'Brien, J. M., Albert, D. M. & Foster, C. S. (1994):
Anterior uveitis.
Principles and practice of ophthalmology. Philadelphia: Saunders, 1, 407-23.

Pearce, J. W., Galle, L. E., Kleiboeker, S. B., Turk, J. R., Schommer, S. K., Dubielizig, R. R., Mitchell, W. J., Moore, C. P. & Giuliano, E. A. (2007):
Detection of Leptospira interrogans DNA and antigen in fixed equine eyes affected with end-stage equine recurrent uveitis.
Journal of Veterinary Diagnostic Investigation, 19 (6), 686-90.

Polle, F., Storey, E., Eades, S., Alt, D., Hornsby, R., Zuerner, R. & Carter, R. (2014):
Role of Intraocular Leptospira Infections in the Pathogenesis of Equine Recurrent Uveitis in the Southern United States.
Journal of Equine Veterinary Science, 34 (11), 1300-6.

Popp, M., Gerhards, H. & Wollanke, B. (2013):
Enrofloxacinkonzentrationen im Glaskörper und Serum von an equiner rezidivierender Uveitis (ERU) erkrankter Pferde nach wiederholter intravenöser Verabreichung.
Pferdeheilkunde, 29 (5), 574-80.

Rebhun, W. C. (1979):
Diagnosis and treatment of equine uveitis.
Journal of the American Veterinary Medical Association, 175 (8), 803-8.

Regan, D. P., Aarnio, M. C., Davis, W. S., Carmichael, K. P., Vandenplas, M. L., Lauderdale, J. D. & Moore, P. A. (2012):
Characterization of cytokines associated with Th17 cells in the eyes of horses with recurrent uveitis.
Veterinary Ophthalmology, 15 (3), 145-52.

Rimpau, W. (1947):
Leptospirose beim Pferd (Periodische Augenentzündung).
Tierärztliche Umschau, 2, 177-8.

Romeike, A., Brügmann, M. & Drommer, W. (1998):
Immunohistochemical studies in equine recurrent uveitis (ERU).
Veterinary Pathology Online, 35 (6), 515-26.

Schwink, K. (1992):
Equine uveitis.
Veterinary Clinics of North America: Equine Practice, 8 (3), 557-74.

Schwink, K., Crisman, M. & Rigg, D. (1989):
Chronic recurrent uveitis in a horse with an elevated aqueous humor antibody titer to Leptospira interrogans serovar autumnalis.
Equine Practice, 11, 41-43.

Sillerud, C. L., Bey, R. F., Ball, M. & Bistner, S. I. (1987):
Serologic correlation of suspected Leptospira interrogans serovar pomona-induced uveitis in a group of horses.
Journal of the American Veterinary Medical Association, 191 (12), 1576-8.

Spiess, B. M. (2010):
Equine recurrent uveitis: the European viewpoint.
Equine Veterinary Journal, (S37), 50-6.

Szemes, P. & Gerhards, H. (2000):
Untersuchungen zur Prävale-nz der equinen rezidivierenden Uveitis im Grossraum Köln-Bonn.
Praktischer Tierarzt, 81 (5), 408-20.

Turner, L. H. (1968):
Leptospirosis. II. Serology.
Transactions of the Royal Society of Tropical Medicine and Hygiene, 62 (6), 880-99.

Tömördy, E., Haessig, M. & Spiess, B. M. (2010):
The outcome of pars plana vitrectomy in horses with equine recurrent uveitis with regard to the presence or absence of intravitreal antibodies against various serovars of Leptospira interrogans.
Pferdeheilkunde, 26 (2), 251-4.

Verma, A., Artiushin, S., Matsunaga, J., Haake, D. A. & Timoney, J. F. (2005):
LruA and LruB, novel lipoproteins of pathogenic Leptospira interrogans associated with equine recurrent uveitis.
Infection and Immunity, 73 (11), 7259-66.

Verma, A., Kumar, P., Babb, K., Timoney, J. F. & Stevenson, B. (2010):
Cross-reactivity of antibodies against leptospiral recurrent uveitis-associated proteins A and B (LruA and LruB) with eye proteins.
PLoS Neglected Tropical Diseases, 4 (8), e778.

Werry, H. & Gerhards, H. (1992):
Zur operativen Therapie der equinen rezidivierenden Uveitis (ERU).
Tierärztliche Praxis, 20 (2), 178-86.

Werry, H. & Gerhards, H. (1991):
Möglichkeiten der und Indikationen zur chirurgischen Behandlung der equinen rezidivierenden Uveitis (ERU).
Pferdeheilkunde, 7 (6), 321-31.

Wiehen, L. E. (2012):
Retrospektive Analyse zum Vorkommen der Equinen rezidivierenden Uveitis -unter Berücksichtigung der Leptospireninfektion- an der LMU München von 01.2005 bis 06.2010 *(Doctorial dissertation, LMU München)*.

Winterberg, A. & Gerhards, H. (1997):
Langzeitergebnisse der Pars-plana-Vitrektomie bei equiner rezidivierender Uveitis. *Pferdeheilkunde*, 13 (4), 377-83.

Wollanke, B. (1995):
Untersuchungen zur Ätiologie der equinen rezidivierenden Uveitis (ERU). (Doctorial dissertation, LMU München).

Wollanke, B. & Gerhards, H. (2009):
Equine rezidivierende Uveitis.
Continuing Veterinary Education, 1 (4), 1-14.

Wollanke, B., Gerhards, H., Brem, S., Kopp, H. & Meyer, P. (1998):
Intraokulare und Serumantikörpertiter gegen Leptospiren bei 150 wegen equiner rezidivierender Uveitis (ERU) vitrektomierten Pferden.
Berliner und Münchener Tierärztliche Wochenschrift, 111, 134-9.

Wollanke, B., Gerhards, H., Brem, S., Meyer, P. & Kopp, H. (2004):
Ätiologie der equinen rezidivierenden Uveitis (ERU): Autoimmunkrankheit oder intraokulare Leptospireninfektion.
Pferdeheilkunde, 20 (4), 327-40.

Wollanke, B., Gerhards, H., Brem, S., Wolf, E., Kopp, H. & Meyer, P. (2000):
Zur Leptospirenätiologie der equinen rezidivierenden Uveitis (ERU): Ergebnisse der Untersuchungen von Serum-und Glaskörperproben.
Tierärztliche Praxis, 28, 153-8.

Wollanke, B., Rohrbach, B. W. & Gerhards, H. (2001):
Serum and vitreous humor antibody titers in and isolation of Leptospira interrogans from horses with recurrent uveitis.
Journal of the American Veterinary Medical Association, 219 (6), 795-800.

10 Danksagung

Ich möchte mich herzlich bei meinem Doktorvater, Herrn Professor Dr. Bernhard Ohnesorge für die Ausarbeitung des Themas und die jederzeit konstruktive Unterstützung während der gesamten Doktorarbeit bedanken.

Frau Dr. Myriam von Borstel möchte ich für die fachliche Unterstützung und Förderung im Bereich der Ophthalmoskopie danken, ohne diese Grundlagen wäre die gesamte Arbeit nicht möglich gewesen.

Bei allen Mitarbeitern der Klinik für Pferde bedanke ich mich für die große Hilfsbereitschaft, insbesondere danken möchte ich Frau Anja Seemann-Jensen und Frau Christiane Richter für die Unterstützung bei der Probenaufbereitung.

Nicht zu vergessen ist die moralische Unterstützung meiner Mitdoktorandin, Frau Christin Schoppe. Doch auch Frau Dr. Silke Ludwig, Frau Dr. Maria Leurs und Frau Katharina Hohlbaum gilt meine Dankbarkeit auf fachlicher, sprachlicher und moralischer Ebene. Ich bin glücklich, dass wir immer aufeinander zählen können.

Weiterhin gilt mein Dank meiner gesamten Familie. Ihr seid der Grundstein meines Lebens, es ist wunderbar, dass ich mich immer auf euch alle verlassen kann. Besonders möchte ich meinem Bruder Herrn Dr. Uwe Nordemann für die jederzeit schnellen Korrekturen danken.

Am Ende bedanke ich mich bei Konstantin, für alles!